Practical Grounding, Bonding, Shielding and Surge Protection

Other titles in the series

Practical Data Acquisition for Instrumentation and Control Systems (John Park, Steve Mackay)

Practical Data Communications for Instrumentation and Control (Steve Mackay, Edwin Wright, John Park)

Practical Digital Signal Processing for Engineers and Technicians (Edmund Lai)

Practical Electrical Network Automation and Communication Systems (Cobus Strauss)

Practical Embedded Controllers (John Park)

Practical Fiber Optics (David Bailey, Edwin Wright)

Practical Industrial Data Networks: Design, Installation and Troubleshooting (Steve Mackay, Edwin Wright, John Park, Deon Reynders)

Practical Industrial Safety, Risk Assessment and Shutdown Systems for Instrumentation and Control (Dave Macdonald)

Practical Modern SCADA Protocols: DNP3, 60870.5 and Related Systems (Gordon Clarke, Deon Reynders)

Practical Radio Engineering and Telemetry for Industry (David Bailey)

Practical SCADA for Industry (David Bailey, Edwin Wright)

Practical TCP/IP and Ethernet Networking (Deon Reynders, Edwin Wright)

Practical Variable Speed Drives and Power Electronics (Malcolm Barnes)

Practical Centrifugal Pumps (Paresh Girdhar and Octo Moniz)

Practical Electrical Equipment and Installations in Hazardous Areas (Geoffrey Bottrill and G. Vijayaraghavan)

Practical E-Manufacturing and Supply Chain Management (Gerhard Greef and Ranjan Ghoshal)

Practical Hazops, Trips and Alarms (David Macdonald)

Practical Industrial Data Communications: Best Practice Techniques (Deon Reynders, Steve Mackay and Edwin Wright)

Practical Machinery Safety (David Macdonald)

Practical Machinery Vibration Analysis and Predictive Maintenance (Cornelius Scheffer and Paresh Girdhar)

Practical Power Distribution for Industry (Jan de Kock and Cobus Strauss)

Practical Process Control for Engineers and Technicians (Wolfgang Altmann)

Practical Telecommunications and Wireless Communications (Edwin Wright and Deon Reynders)

Practical Troubleshooting Electrical Equipment (Mark Brown, Jawahar Rawtani and Dinesh Patil)

Practical Grounding, Bonding, Shielding and Surge Protection

G. Vijayaraghavan, B.Eng (Hons) Consulting Engineer, Chennai, India

Mark Brown, Pr.Eng, DipEE, B.Sc (Elec.Eng), Senior Staff Engineer, IDC Technologies, Perth, Australia.

Malcolm Barnes, CPEng, BSc (ElecEng), MSEE, Alliance Automation, Perth, Western Australia

Series editor: Steve Mackay

ELSEVIER

AMSTERDAM • BOSTON • HEIDELBERG • LONDON
NEW YORK • OXFORD • PARIS • SAN DIEGO
SAN FRANCISCO • SINGAPORE • SYDNEY • TOKYO
Newnes is an imprint of Elsevier

Newnes

Newnes
An imprint of Elsevier
Linacre House, Jordan Hill, Oxford OX2 8DP
200 Wheeler Road, Burlington, MA 01803

First published 2004

British Library Cataloguing in Publication Data
Vijayaraghavan, G.
    Practical grounding, bonding, shielding and surge
    protection. – (Practical professional)
    1. Electric apparatus and appliances – Protection   2. Electric currents – Grounding
    I. Title
    621. 3'17

Library of Congress Cataloguing in Publication Data
A catalogue record for this book is available from the Library of Congress

ISBN   0 7506 6399 5

For information on all Newnes publications
visit our website at www.newnespress.com

Typeset and edited by Integra Software Services Pvt. Ltd, Pondicherry, India
www.integra-india.com

Working together to grow
libraries in developing countries

www.elsevier.com | www.bookaid.org | www.sabre.org

ELSEVIER   BOOK AID International   Sabre Foundation

Transferred to digital printing 2007

Contents

Preface

Few topics generate as much controversy and argument as that of grounding and the associated topics of surge protection, shielding and lightning protection of electrical and electronic systems. Poor grounding practice can be the cause of continual and intermittent difficult-to-diagnose problems in a facility. This book looks at these issues from a fresh yet practical perspective and enables you to reduce expensive downtime on your plant and equipment to a minimum by correct application of these principles. This book is designed to demystify the subject of grounding and presents the subject in a clear, straightforward manner. Installation, testing and inspection procedures for industrial and commercial power systems will be examined in detail. Essentially the discussion in this book is broken down into grounding, shielding and surge protection for both power and electronics systems. Grounding and surge protection for Telecommunications and IT systems are examined in detail. Finally, the impact of lightning is examined and simple techniques for minimizing its impact are described. The terms grounding and earthing are understood to be interchangeable in this book but due to the larger readership the term grounding has been the preferred usage. Our apologies to our European readers for this unfortunate compromise.

Typical people who will find this book useful include:

- Instrumentation and Control Engineers
- Consulting Engineers
- Electrical Engineers
- Project Engineers
- Maintenance Engineers
- Electrical Contractors
- Safety Professionals
- Consulting Engineers
- Electricians
- Electrical Inspectors
- Power System Protection and Control Engineers
- Building Service Designers
- Data Systems Planners and Managers
- Electrical and Instrumentation Technicians.

We would hope that you will gain the following from this book:

- Knowledge of the various methods of grounding electrical systems
- Details of the applicable national Standards
- The purposes of grounding and bonding
- A list of the types of systems that cannot be grounded
- Details on how to correctly shield sensitive communications cables from noise and interference
- Know-how on surge and transient protection

- The ability to troubleshoot and fix grounding and surge problems
- A good understanding of lightning and how to minimize its impact on your facility.

Some working knowledge of basic electrical engineering principles is required, although there will be a revision at the beginning of the book. Experience with grounding problems will enable the book to be placed in context.

1

Introduction and basics

1.1   Introduction

The practice of grounding of electrical systems is almost as old as the development and widespread use of electric power itself. In this book, we will take a look at the need for adopting good grounding practices at both the source of power (a generator or a transformer substation) and the consumer premises. We will study various methods of grounding of electrical systems and make a comparison of their effectiveness. We will learn about electric shock and how to prevent electrical accidents by timely detection and isolation of faulty equipment.

We will discuss the effect of lightning on electrical systems and the means of protecting the systems from damage by safely conducting away the surges caused by lightning strokes into ground. We will learn the method of establishing reliable ground connections, to predict by calculating the earth resistance and the methods for measurement of earth resistance of grounding systems.

We will also review why even non-electrical gear parts of certain types of machinery will have to be connected or bonded to ground to prevent accumulation of static charges, which would otherwise cause sudden and destructive spark-over.

We will also review the practices adopted for grounding and bonding in consumer premises and their importance in modern day systems with a lot of sensitive electronic equipment (which create as well as are affected by phenomena such as surges, electrical noise, etc.). Further, we will detail the importance of shielding of signal wires and establishing a zero signal reference grid in data processing centers. We will study the generation of harmonics and how they affect electrical equipment, as well as the means to avoid them.

We will learn about power quality and the role of uninterrupted power supply (UPS) systems in overcoming some of the power supply problems and discuss various possible configurations of static UPS systems and the issues pertaining to the grounding of UPS fed systems.

Note: The terms *earth* and *ground* have both been in general use to describe the common signal/power reference point and have been used interchangeably around the world in the electrotechnical terminology. The IEEE Green Book, however, presents a convincing argument for the use of the term *ground* in preference to *earth*. An electrical ground need not necessarily be anywhere near the earth (meaning soil). For a person working in the top floor of a high-rise building, electrical ground is far above the earth. In deference to this argument, we will adopt the term ground in this manual to denote the common electrical reference point.

1.2 Basics of grounding

Grounding serves the following principal purposes:

- It provides an electrical supply system with an electrical reference to the groundmass. By connecting a particular point of the supply source to the ground (such as the neutral of a three-phase source), it is ensured that any other point of the system stays at a certain potential with reference to the ground.
- A metallic surface of the enclosure of an electrical system is grounded to ensure that it stays at ground potential always and thus remains safe to persons who may come into contact with it.
- It provides a low-impedance path for accumulated static charges and surges caused by atmospheric or electrical phenomenon to the ground thus ensuring that no damage is caused to sensitive equipment and personnel.

Electrical systems were not always grounded. The first systems were ungrounded ones with no ground reference at all. Even though such systems still exist in specific areas, they are the exceptions rather than the rule and by and large, some form of grounding is adopted for all power systems. We all know that the insulating layer around the current-carrying conductors in electrical systems is prone to deterioration. When a failure of insulation takes place due to aging, external factors or due to electrical or thermal stress, it is necessary to detect the point of failure so that repairs can be undertaken. In a system that has no ground reference at all, it is not easy to correctly pinpoint the faulted location. Refer to Figure 1.1a, which shows such a system. It can be seen that due to the absence of a conducting path through ground, the fault remains undetected. If, however, a second fault occurs in the unaffected line at some other point in the system, it can cause a shorting path and results in the flow of high magnitude fault currents that can be detected by protective devices.

To detect the first fault point as soon as it happens without waiting for a second fault to develop, we ground one of the two poles of the source S (refer Figure 1.1b). The pole that is grounded is generally called the neutral and the other, 'line'. It would be of interest to note that the connection between neutral and earth is only at the source. The return current from the load flows only through the neutral conductor back to the source. For this reason, the neutral is always insulated from ground and usually to the same degree as the line conductor. When there is an insulation failure in the line conductor, a high current flows through the electrical circuits and through the ground path back to the source and depending on the resistance of the ground path, the current flow in this path can be detected by appropriate protective equipment.

Thus, one of the primary purposes of grounding is to permit easy detection of faults in electrical systems by providing a path for the flow of currents from the fault point through the ground (and sometimes the earth mass) back to the neutral point of the source.

Now let us take a step further and see as to why it is necessary for this ground reference to be extended to the consumer installation. While Figure 1.1b shows that the source is grounded, it does not indicate another point of connection to ground. However, in practical systems, the fact that a failure of insulation takes place does not mean that a ground connection is automatically established. This can only be done if the point of failure is connected to ground through a low-impedance ground path. Such a path is created using a reference ground bus at the consumer end and connecting the metallic housing of all electrical equipment to this bus (refer Figure 1.2).

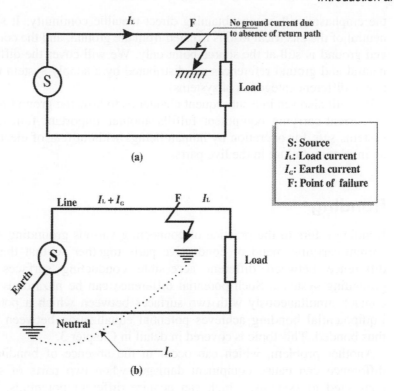

Figure 1.1
(a) Fault in ungrounded system, (b) Effect of grounding the neutral

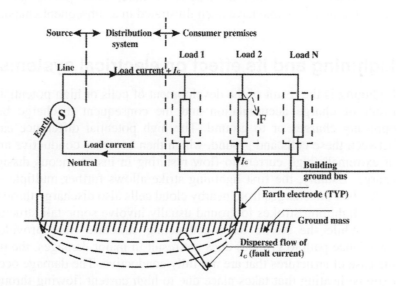

Figure 1.2
Fault current flow in a grounded system

In fact, it is preferable to have the ground terminal of a low voltage consumer installation directly connected to the neutral of the source to ensure that the ground fault current has a low-impedance path not involving the earth mass. It is difficult to predict accurately the resistance of groundmass to the flow of currents and hence except for high voltage systems, the emphasis will be on obtaining direct metallic continuity. It should be noted that the neutral of the electrical load is isolated from the ground, and the connection between neutral and ground is still at the source point only. We will cover the different ways in which the neutral and ground references are distributed by a supply system to its consumers (giving rise to different categories of systems).

We will also see in a subsequent chapter as to how the grounding of metallic enclosures of current-carrying equipment fulfills another important function: that of making the systems safe for operation by human beings without fear of electrocution in the event of an insulation failure in the live parts.

1.3 Bonding

Bonding refers to the practice of connecting various grounding systems as well as non-current-carrying metal or conductive parts together so that there will be no potential difference between different accessible conducting surfaces or between different grounding systems. Such potential difference can be hazardous if a person comes into contact simultaneously with two surfaces between which a potential difference exists. Equipotential bonding achieves potential equalization between all surfaces, which are thus bonded. This topic is covered in detail in Chapter 3.

Another problem, which can occur in the absence of bonding, is that the potential difference can cause equipment damages when two parts of sensitive equipment are connected to systems, which can acquire different potentials. The currents that flow through inter-system capacitances can cause damage to sensitive components and printed circuit boards. This type of problem generally occurs when ground current surges happen as a result of lightning discharges or other atmospheric phenomena. Case studies involving this principle have been illustrated in a subsequent chapter.

1.4 Lightning and its effect on electrical systems

Lightning is the result of the development of cells of high potential in cloud systems as a result of charge accumulation and the consequent discharge between cells carrying opposing charges or to ground. The high potential difference causes ionization of air between these cells and ground, which then becomes conductive and allows a short burst of extremely high current to flow resulting in instantaneous dissipation of accumulated charge. Usually, the first lightning strike allows further multiple strikes along the same path when the charges from nearby cloud cells also discharge through it to ground.

The lightning strokes to ground usually involve some tall structure or object such as a tree. While the stroke on a conducting structure (that provides an extremely low-impedance path to ground) does not result in major damages, the results are disastrous in the case of structures that are not fully conductive. The damage occurs mainly because of extreme heating that takes place due to high current flowing through the object. This, in turn, causes any moisture present in the structure to evaporate suddenly. The resulting explosive release of steam causes extensive damage to the object. For example, in a tree that suffers a lightning stroke, the moist layer under its bark vaporizes instantaneously which causes the bark to fly away.

A more serious result can occur if the stroke occurs near or on a container carrying flammable materials. The high temperatures can ignite the flammable materials causing severe explosions and secondary damages. Such structures need special protection against lightning strokes.

Of greater interest to us in this book is the effect lightning discharge has on electrical systems and how electrical equipment and installations can be protected against damage. These will be dealt in detail in a subsequent chapter.

1.5 Static charges and the need for bonding

Certain types of non-electrical machinery can cause a buildup of static charge during their operation and this charge accumulates on the surface of the equipment parts (for example, a flat rubber belt around two metal pulleys, which is a very common type of motive power transmission, generates a lot of static electricity). When a sufficient amount of charge is built up, a spark-over can occur between the charged part and any grounded body nearby. Figure 1.3 illustrates the principles involved.

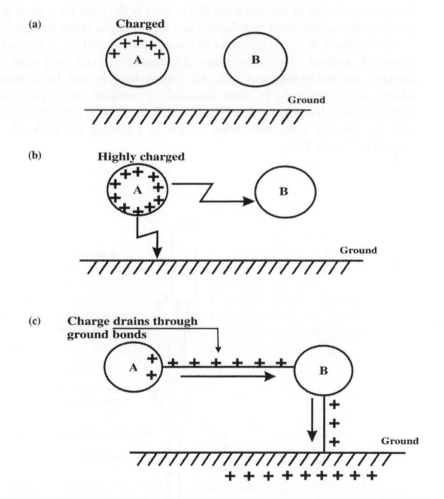

Figure 1.3
Example of static electricity buildup and prevention

Body A has a positive charge while no charge is present on the nearby body B, both of which are insulated from the ground (Figure 1.3a). Let us now assume that body B is connected with ground. When body A acquires a sufficient quantum of charge that can cause breakdown of the medium separating A and B or A and ground, it will result in a spark discharge (Figure 1.3b).

Such spark-over carries sufficient energy that can cause explosions in hazardous environments and fires in case combustible materials are involved. It is therefore necessary to provide bonding of the parts where charge buildup can occur by suitable metallic connections to earth. Bonding bodies A and B with a conducting metallic wire causes the charge to flow on to body B. This causes the charge to continuously leak into the ground so that buildup of dangerously high voltages is prevented (refer to Figure 1.3c).

Some of the practical cases of static buildup that occur in industrial and consumer installations and ways and means of avoidance will be dealt in further detail in a subsequent chapter.

1.6   Ground electrodes and factors affecting their efficacy

A common thread in the foregoing discussions is the need for a good ground connection in power sources, consumer installations and for structures prone to lightning strokes.

The connection to groundmass is normally achieved by a ground electrode. Several types of ground electrodes using different materials, physical configurations and designs are in widespread use and follow usually the local standards that govern electrical installations. In most standards, a metallic rod driven into the ground to a depth where adequate moisture is available in the soil throughout the year in both wet and dry seasons is recommended for use as a ground electrode. A typical electrode is shown in Figure 1.4.

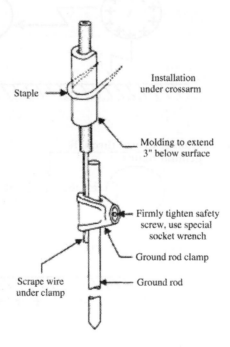

Figure 1.4
A typical ground electrode used in electrical installations

The performance of such electrodes (considering the ground resistance of the electrode as an indicator) depends on the type of soil, its composition, conductivity, presence of moisture, soil temperature, etc. Several ground electrodes bonded together to form a cluster are usually provided for achieving satisfactory results. The general requirements that influence the choice of earth electrodes are as follows:

- The type of soil where the grounding is carried out (in particular, its electrical resistivity).
- The need for achieving minimum acceptable earth resistance appropriate to the installation involved.
- The need to maintain this resistance all round the year in varying climatic conditions.
- Presence of agents that can cause corrosion of elements buried in ground.

The electrode design and methods of installation will be dependent on these requirements. These will be taken up in detail in a later chapter.

To improve the conductivity of ground electrodes, several forms of electrode construction are in use in which the layer of soil surrounding the electrode is treated with chemical substances for improving the conductivity. These are known as chemical electrodes. The basic principle of these electrodes is the use of substances that absorb moisture and retain it over long periods. These are packed as backfill around the electrode. Materials containing carbon (charcoal/coke) and electrolyte salts such as sodium chloride are typically used as backfill. Figure 1.5 shows such an electrode construction. It may also be noted that in this construction, a provision has been made to add water externally to keep the backfill material wet during prolonged dry weather conditions.

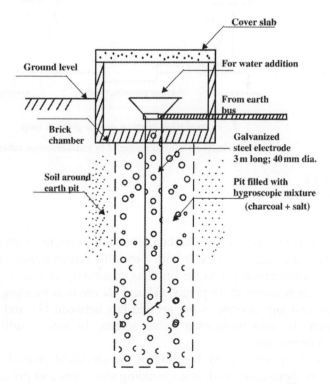

Figure 1.5
A typical chemical ground electrode

It will also be evident from the above discussions that being a critical factor in the safety of installation and personnel, the grounding system will have to be constantly monitored to ensure that its characteristics do not drift beyond acceptable limits. The practical methods adopted for measurement of soil resistivity and the resistance of a ground electrode/grounding system will also be covered in later chapters.

1.7   Noise in signaling circuits and protective measures such as shielding

Noise in communication and signaling circuits is an issue needing careful consideration while planning and installing a system. Noise can be due to improper earthing practices. It may also arise from surges due to external or internal causes (explained in the previous section) and by interference from other nearby circuits. Incorrect earthing can result in noise due to ground loops. Figure 1.6 is an example of such a situation.

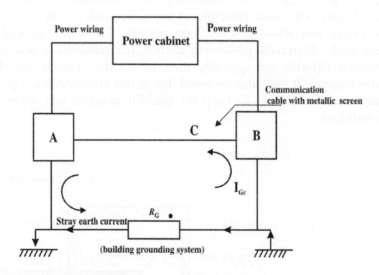

I_{Gc}: circulating current due to ground loop

A, B: electronic cabinets connected by communication cable

Figure 1.6
Ground loop problem

A and B are two electronic data processing systems with a communication connection C between them. C is a cable with a metallic screen bonded to the enclosures of A and B. A and B are grounded to the building grounding system at points G1 and G2. R_G is the resistance between these points. G1 and G2 are thus forming a ground loop with the cable screen and any current in the ground bus between G1 and G2 causes a current to flow through the communication cable screen, in turn resulting in spurious signals and therefore malfunction.

Multiple grounding by bonding of the electrical ground wire to the conduits and the conduits themselves with other building structures and piping is done in electrical wiring to get a low ground impedance and it has no adverse effect on power electrical devices. In fact, many codes recommend such practices in the interest of human safety. However, as shown in the above example, the same practice can cause problems when applied to noise-sensitive electronic equipment.

Noise can be due to galvanic coupling, electrostatic coupling, electromagnetic induction or by radio frequency interference. Both normal signals and surge as well as power frequency currents can affect nearby circuits with which, they have a coupling. Design of certain types of signal connections has an inherent problem of galvanic coupling. Electrostatic coupling is unavoidable due to the prevalence of inter-electrode capacitances especially in systems handling high-frequency currents. Most power electrical equipment produce electromagnetic fields. Arcing in the contacts of a switching device producing electromagnetic radiation or high-frequency components in currents flowing in a circuit setting up magnetic fields when passing through wiring are examples of such disturbances. This kind of disturbance is called electromagnetic interference (EMI). By and large, the equipment being designed these days have to conform to standards, which aim to reduce the propagation of EMI as well as mitigating the effects of EMI from nearby equipment by using appropriate shielding techniques. Shielding against electrostatic coupling and electromagnetic interference works differently and should be applied depending on the requirements of the given situation.

The method of grounding the electromagnetic and electrostatic shield/screen is also important from the point of view of noise. Improperly grounded shield/screen can introduce noise into signaling and communication systems, which it is to protect.

Since electronic equipment involve the use of high-frequency signals, the impedance of the grounding system (as against resistance, which we normally consider) assumes significance. The ground system design for such equipment must take the impedance aspect into consideration too.

These and other common problems faced in the electrical systems of today will be dealt in greater detail in subsequent chapters. Remedial measures to avoid such phenomena from affecting sensitive circuits will also be discussed.

We will also briefly touch upon the subject of harmonics, which are sometimes a source of noise. Harmonics are voltage/current waveforms of frequencies, which are multiples of the power frequency. Harmonics are generated when certain loads connected to the system draw currents that are not purely sinusoidal in waveform (such non-sinusoidal current waveforms can be resolved into a number of sinusoidal waveforms of the fundamental power frequency and its multiples). Many of the modern devices using semiconductor components belong to this type and constitute what are called non-linear loads. Harmonic current being of higher frequencies causes audible hum in communication circuits and can interfere with low amplitude signals. They also cause heating in equipment due to higher magnetic loss and failure of capacitor banks due to higher than normal current flow. We will cover the basic principles briefly in this book.

1.8   Surge protection of electronic equipment

Modern day industries and businesses rely largely on electronic systems for their smooth functioning, be it industrial drives, distributed control systems, computer systems and networking equipment or communication electronics. These electronic devices often work with very low power and voltage levels for their control and communications and cannot tolerate even small over-voltages or currents. Induced voltages from nearby power circuits experiencing harmonic current flow can also cause interference in the systems carrying communication signals and can result in malfunctions due to erroneous or noisy signal transmission. Due to this sensitive nature of electronic and communication equipment, any facility that houses such equipment needs to have its electrical wiring and grounding systems planned with utmost care so that there are no unpredictable equipment failures or malfunction.

Another problem is that of voltage spikes that occur in the power supply. Some of these may originate from the external grid but some others may originate from other circuits within the same premises. The result of such voltage/power surges is invariably the failure of the electronic device itself. A typical example of an external voltage disturbance is a lightning stroke near an overhead power transmission system. Such transients can also happen due to switching on or off large transformers. The transformers when charged draw a momentary inrush current and this can reflect as a voltage disturbance. Similarly, switching off an inductive load (say a coil energizing a contactor) causes a brief voltage spike due to the collapse of the magnetic field in the magnetic core. If other equipment are connected in parallel with the inductance (after the switching point), they will experience the surge. Figure 1.7 shows the principle involved.

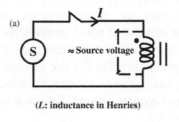

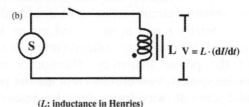

Figure 1.7
Inductive load causing a transient surge

Any installation can be divided into various zones depending on the severity of surges to which equipment in the zone can be subjected. Surge protective devices and grounding are arranged in such a way that surge levels gradually get reduced from the most severe magnitudes in zone 0 to the highly protected zone 3, which houses the most sensitive and vulnerable systems. This is explained in Chapter 7 in detail. Different types of surge protective devices available and their application areas will also be touched upon.

1.9   UPS systems and their role in power quality improvement

Widespread use of process control/SCADA systems in industries and computers and communication equipment in business environment demands an uninterrupted power supply of good quality free from harmonics, voltage irregularities, etc. since these systems are sensitive to power interruptions and voltage/frequency excursions. Various uninterrupted power supply options of both electromechanical and static variety are available in the market. The electromechanical type of systems usually has a prime mover such as an engine-driven alternator. These systems have some form of intermediate energy storage, which permits the alternator to ride through power system disturbances and also provide the energy required to start the engine.

However, the static invertor-fed UPS systems deriving standby power from a storage battery are the most commonly used type of system. In the present day systems, electronic UPS systems, switching elements of insulated gate bipolar transistors (IGBT) are employed to advantage. Besides making the switching circuit simple, they also permit the invertors to switch at very high frequencies. This in turn reduces noise, makes the systems more compact and with pulse width modulation techniques ensures a harmonic-free output.

There are, however, a number of factors that come into play while selecting small and medium capacity UPS systems. This will be discussed in the concluding chapter of this book. The issues relating to the grounding of these systems, as separately derived sources, have been extensively discussed in IEEE: 142 publications. We will discuss some of the commonly used UPS configurations in detail.

1.10   Case studies

This book contains several facts that need attention in designing and implementing electrical and instrumentation systems. In order to illustrate the problems, which can be caused by not paying adequate attention to these aspects, a chapter outlining several case studies has been included. These case studies cover the problems encountered in various real life situations and have been illustrated in order to highlight the principles that have been dealt with in this book.

1.11   Importance of local codes

While we covered, in the above discussion, the general physical principles involved in grounding of electrical systems and equipment, the actual practices adopted in different countries may vary. Different local codes approach the issue of grounding in their own different ways, taking into account factors such as the local environment, material availability and so on. All of them, however, aim to achieve certain common objectives, most important of them being the safety of personnel. Generation, distribution and use of electrical energy are subject to extensive regulatory requirements of each country such as the National Electrical Code (USA) and the Wiring Regulations (UK). It is mandatory on the part of electricity suppliers and consumers to adopt the practices stipulated in these regulations and any deviations might cause the installation in question to be refused permission for operation. Thus, anyone engaged in the planning and design of electrical systems must be well versed with the applicable local codes and must take adequate care to ensure conformity to the codes in all mandatory aspects.

1.12   Summary

In this chapter, we had a broad overview of the need for grounding of electrical systems and the practices adopted for grounding. We covered the topics of lightning and static charges and learnt that precautions are needed to mitigate their effects. We discussed the configuration of a typical ground electrode and their general requirements. We had touched briefly on the need for proper protection of sensitive equipment to phenomena such as surges, harmonics and EMI. We discussed how inadvertent ground loops could affect the operation of sensitive signal circuits. We also reviewed the role of UPS systems in ensuring proper power quality. With this background information, we will proceed for a more in-depth discussion on these topics in the coming chapters.

2

Grounding of power supply system neutral

2.1   Introduction

As we had seen in the previous chapter, grounding of supply system neutral fulfills two important functions.

1. It provides a reference for the entire power system to the groundmass and establishes a path for flow of currents to ground whenever there is a failure of insulation so that the fault can be detected by circuit protective devices and isolated.
2. It ensures that in the event of an accidental connection of live parts to a conducting metallic enclosure, any person coming into contact with the enclosure does not experience dangerously high voltages. This is done by bonding the enclosure to the ground so that the enclosure's potential is firmly 'clamped' to that of the ground. Also, bonding of all exposed metal parts in a building and connecting them to ground creates an equipotential environment where all such parts will be essentially at the same potential as the ground.

In this chapter, we will learn about the various types of grounding an electrical system and their relative advantages. As you may recall from the previous chapter, grounding of both source and the consumer equipment is necessary. What we will see in this chapter is about the grounding of the power source.

Note: We will be discussing in this as well as in the subsequent chapters electrical systems of three-phase configuration since for all practical purposes, this is the only configuration that utilities all over the world adopt. Systems of single-phase configuration will, however, be used in illustrations for simplicity. Figure 2.1 shows the various types of grounding methods that are possible.

The diagrammatic representation of these different grounding techniques and the equivalent impedances are shown in Figure 2.2. We will go through in detail about each method in the subsequent paragraphs.

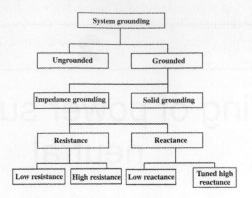

Figure 2.1
Grounding methods

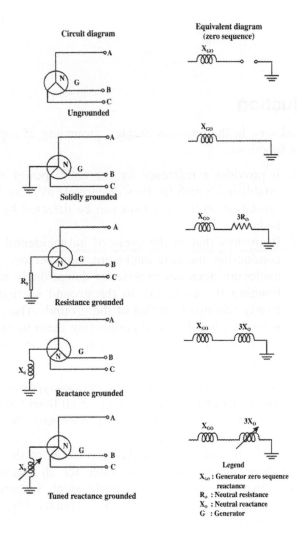

Figure 2.2
Grounding techniques and equivalent impedances

2.2   **Ungrounded systems**

As discussed in Chapter 1, providing a reference ground in an electrical system is essential for safe operation. But there are certain cases in which a system can be operated without such a reference.

By definition, an electrical system, which is not intentionally connected to the ground at any point, is an ungrounded system. However, it should be noted that a connection to ground of sort does exist due to the presence of capacitances between the live conductors and ground, which provides a reference. But these capacitive reactances are so high that they cannot provide a reliable reference. Figure 2.3 illustrates this point. In some cases, the neutral of potential transformer primary windings connected to the system is grounded, thus giving a ground reference to the system.

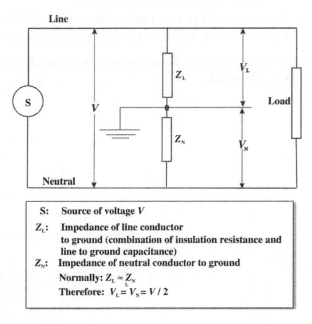

Figure 2.3
A virtual ground in an ungrounded system

It may be noted that normally the capacitance values being equal to the lines L1 and L2 are roughly at a potential equal to half the voltage of the source from the ground (it is possible to demonstrate this by measurement of a high-impedance device such as an electrostatic type of voltmeter).

The main advantage cited for ungrounded systems is that when there is a fault in the system involving ground, the resulting currents are so low that they do not pose an immediate problem to the system. Therefore, the system can continue without interruption, which could be important when an outage will be expensive in terms of lost production or can give rise to life-threatening emergencies.

The second advantage is that one need not invest on elaborate protective equipment as well as grounding systems, thus reducing the overall cost of the system. (In practice, this is however offset somewhat by the higher insulation ratings which this kind of system calls for due to practical considerations.)

The disadvantages of such systems are as follows:

- In all but very small electrical systems, the capacitances, which exist between the system conductors and the ground, can result in the flow of capacitive current at the faulted point which can cause repeated arcing and buildup of excessive voltage with reference to ground. This is far more destructive and can cause multiple insulation failures in the system at the same instant.
- The second disadvantage in practical systems is that of detecting the exact location of the fault, which could take far more time than with grounded systems. This is because the detection of fault is usually done by means of a broken delta connection in the voltage transformer circuit (Figures 2.4a and b). This arrangement does not tell where a fault has occurred and to do so, a far more complex system of ground fault protection is required which negates the cost advantage we originally talked about.
- Also, a second ground fault occurring in a different phase when one unresolved fault is present, will result in a short circuit in the system.

Due to these overwhelming disadvantages, very rarely, if ever, distribution systems are operated as ungrounded.

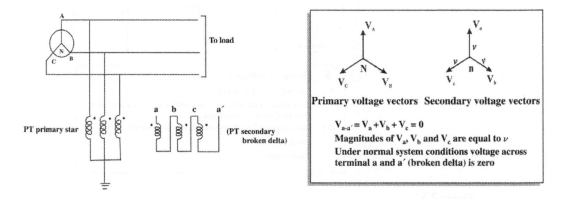

Figure 2.4a
Detection of ground fault using a broken delta connection – under normal condition

2.3   Solidly grounded systems

As is evident from the name, a solidly grounded system is one where the neutral of the system is directly connected to ground without introducing any intentional resistance in the ground circuit. With appropriate choice of the type and number of grounding electrodes, it is possible to obtain a very low-impedance ground connection, sometimes as low as 1 Ω.

A solidly grounded system clamps the neutral tightly to ground and ensures that when there is a ground fault in one phase, the voltage of the healthy phases with reference to ground does not increase to values appreciably higher than the value under the normal operating conditions.

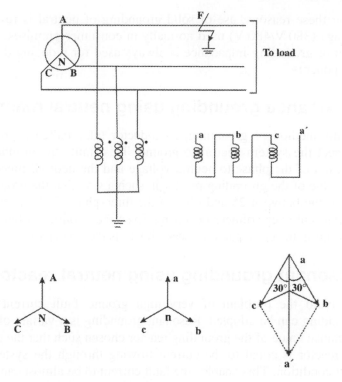

When there is an Earth fault in line A it assumes Earth Potential.

Therefore Voltage across PT primary windings become

$$V_A = 0, \ V_B = V_{BA.} \ , V_C = V_{CA}$$

Thus Secondary Vectors are

$$V_a = 0, \ V_b = V_{ba.} \ , V_c = V_{ca}$$

Figure 2.4b
Detection of ground fault using a broken delta connection – under Ground Fault condition

The advantages of this system are:

- A fault is readily detected and therefore isolated quickly by circuit protective devices. Quite often, the protection against short circuit faults (such as circuit breakers or fuses) is adequate to sense and isolate ground faults as well.
- It is easy to identify and selectively trip the faulted circuit so that power to the other circuits or consumers can continue unaffected (contrast this with the ungrounded system where a system may have to be extensively disturbed to enable detection of the faulty circuit).
- No possibility of transient over-voltages.

The main disadvantage is that when applied in distribution circuits of higher voltage (5 kV and above), the very low ground impedance results in extremely high fault currents almost equal to or in some cases higher than the system's three-phase short circuit currents. This can increase the rupturing duty ratings of the equipment to be selected in these systems.

Such high currents may not have serious consequences if the failure happens in the distribution conductors (overhead or cable). But when a fault happens inside a device such as a motor or generator such currents will result in extensive damage to active magnetic parts through which they flow to reach the ground.

For these reasons, use of solid grounding of neutral is restricted to systems of lower voltage (380 V/480 V) used normally in consumer premises. In all the other cases, some form of grounding impedance is always used for reducing damage to critical equipment components.

2.4　Impedance grounding using neutral reactor

In this method of grounding, an inductor (also called a grounding reactor) is used to connect the system neutral to ground. This limits the ground fault current since it is a function of the phase to neutral voltage and the neutral impedance. It is usual to choose the value of the grounding reactor in such a way that the ground fault current is restricted to a value between 25 and 60% of the three-phase fault current to prevent the possibility of transient over-voltages occurring. Even these values of fault current are high if damage prevention to active parts (as seen above) is the objective.

2.5　Resonant grounding using neutral reactor

To avoid the problem of very high ground fault currents, the method of resonant grounding can be adopted. Resonant grounding is a variant of reactor grounding with the reactance value of the grounding reactor chosen such that the ground fault current through the reactor is equal to the current flowing through the system capacitances under such fault condition. This enables the fault current to be almost canceled out resulting in a very low magnitude of current, which is in phase with the voltage. This serves the objectives of low ground fault current as well as avoiding arcing (capacitive) faults, which are the cause of transient over-voltages. The action is explained in Figure 2.5.

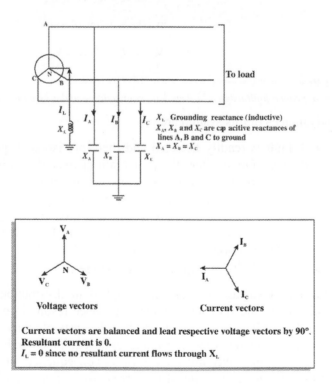

Figure 2.5
Resonant grounding

This type of grounding is common in systems of 15 kV (primary distribution) range with mainly overhead lines but is not used in industrial systems where the reactor tuning can get disturbed due to system configuration changes caused by switching on or off cable feeders (with high capacitive currents) frequently.

2.6   Impedance grounding through neutral resistance

This is by far the most common type of grounding method adopted in medium voltage circuits. The system is grounded by a resistor connected between the neutral point and ground. The advantages of this type of grounding are as follows:

- Reducing damage to active magnetic components by reducing the fault current.
- Minimizing the fault energy so that the flash or arc blast effects are minimal thus ensuring safety of personnel near the fault point.
- Avoiding transient over-voltages and the resulting secondary failures.
- Reducing momentary voltage dips, which can be caused if, the fault currents were higher as in the case of a solidly grounded system.
- Obtaining sufficient fault current flow to permit easy detection and isolation of faulted circuits.

Resistance grounding can again be sub-divided into two categories, viz. high-resistance grounding and low-resistance grounding.

High-resistance grounding limits the current to about 10 A. But to ensure that transient over-voltages do not occur, this value should be more than the current through system capacitance to ground. As such, the applications for high-resistance grounding are somewhat limited to cases with very low tolerance to higher ground fault currents. A typical case is that of large turbine generators, which are directly connected to a high-voltage transmission system through a step up transformer. The capacitance current in generator circuits is usually very low permitting values of ground fault currents to be as low as 10 A. The low current ensures minimal damage to generator magnetic core thus avoiding expensive factory repairs. Figure 2.6 illustrates a practical case of grounding the neutral of a generator of this type.

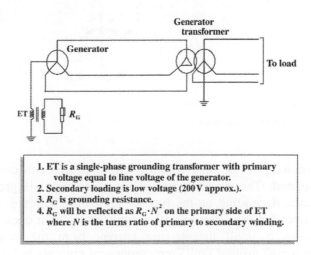

Figure 2.6

Grounding of a turbine generator neutral through a high neutral resistance

On the other hand, a low-resistance grounding is designed for ground fault currents of 100 A or more with values of even 1000 A being common. The value of ground fault current is still far lower than three-phase system fault currents. This method is most commonly used in industrial systems and has all the advantages of transient limitation, easy detection and limiting severe arc or flash damages from happening.

2.7   Point of grounding

In most three-phase systems, the neutral point at source (a generator or transformer) is connected to ground. This has the advantage of minimum potential of the live terminals with reference to ground.

In the case of generators, which are almost always star (wye) connected, the neutral point is available for grounding. However, in the case of transformer substations, a neutral may not always be available as the winding may be delta connected. In such cases, it will be necessary to obtain a virtual neutral using a device called grounding transformer.

Grounding transformers are generally of two types viz. zig-zag connected transformer with no secondary winding and a wye-delta transformer. Figure 2.7 shows a zig-zag grounding transformer.

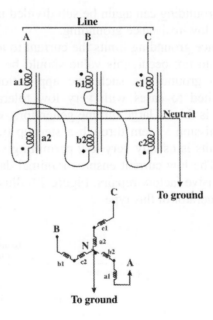

Figure 2.7
Zig-zag grounding transformer

The transformer primary winding terminals are connected to the system, which has to be grounded. The neutral point of the transformer is grounded solidly or through an impedance depending on the type of grounding selected. Under normal conditions, the transformer behaves like any other transformer with open circuited secondary (no-load) and draws a small magnetizing current from the system. The impedance of the transformer to ground fault (zero sequence) currents is however extremely small. When one of the lines develops a ground fault, the current is only restricted by the grounding impedance. Thus, the system behaves virtually in the same manner as any system with grounded source neutral. Figure 2.8 shows this behavior. The ground fault current flowing in the faulted line divides itself into three equal parts flowing through each phase winding of the transformer.

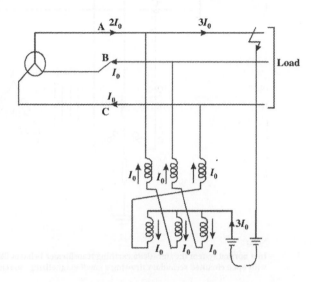

Earthing transformer acts like a low impedance for zero sequence currents

Figure 2.8
Behavior of a zig-zag connected transformer during a ground fault

The other type of grounding transformer is a wye-delta connected transformer. The primary winding terminals of the transformer are connected to the system, which is to be grounded, the neutral of the primary is connected to the ground and the secondary delta is either kept open or can be connected to a three-phase three-wire supply system as required (refer to Figure 2.9).

This type of transformer too presents a low-impedance path to the flow of zero sequence currents due to the circulating path offered by the secondary delta winding. This enables the ground fault current to flow through the primary and to the ground through the grounding impedance. Figure 2.10 illustrates this action.

British Standard BS: 7671:2000 (IEE Wiring regulations) discusses the grounding of low-voltage installations in detail and has provided a method of classifying supply systems based on the type of grounding adopted as well as the method used to extend the system ground to consumer installations. The standard also discusses the comparative merits of the different types of systems for specific applications (refer to Appendix A for details of this classification).

2.8 Other challenges

In the above discussions, we dealt with systems having a single source. However, when more than one source is involved (such as multiple generators or a mix of generators and transformers), grounding of neutrals becomes even more of a challenge. The guiding principles are still the same, viz. the need for limiting the fault current to safe but easily detectable values and the prevention of transient over-voltages during a ground fault.

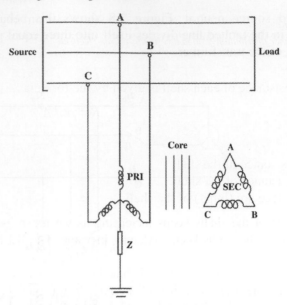

Note:

In a normal system the star–delta earthing transformer behaves like a transformer with open circuited secondary drawing a small magnetizing current from the system.

Figure 2.9

Star–delta grounding transformer

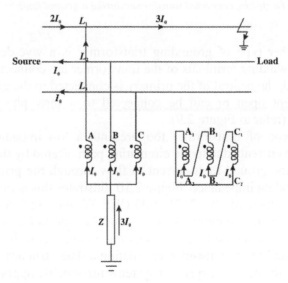

Note:

When there is an earth fault in any line, star–delta transforner acts as a low-impedance path for zero sequence current.

Figure 2.10

Behavior of star–delta grounding transformer during system ground faults

In the case of power distribution systems with several voltage levels separated by transformers, it is necessary to establish neutral grounding for each individual system, taking into consideration the principles cited above and the characteristics of each system.

Neutral grounding of electrical systems within large mobile equipment having their own step down transformers presents further complexities. These are however beyond the scope of this book and are not therefore elaborated.

2.9   Summary

In this chapter, we discussed various types of grounding electrical systems and the considerations that govern the choice. We have also seen how a virtual neutral point can be established through indirect means using a grounding transformer. Different types of supply systems based on neutral wiring practices were reviewed.

3

Equipment grounding

3.1　Introduction

The previous chapter dealt with the grounding of neutral point of electrical systems at the source of power. In this chapter we will learn about the whys and hows of grounding of electrical equipment at the point of utilization.

The basic objectives of grounding of electrical equipment enclosures are as follows:

- To reduce electric shock hazards to personnel
- To provide a low-impedance return path for ground fault currents to the power source so that the occurrence of fault can be sensed by the circuit protective devices and faulty circuit can be safely isolated
- To minimize fire or explosion hazard by providing a ground path of adequate rating, matching the let through energy by circuit protective devices
- To provide a path for conducting away leakage current (small currents flowing through electrical protective insulation around live conductors) and for accumulated static charges (covered in a later chapter).

We will review each of these functions in the subsequent paragraphs.

3.2　Shock hazard

The human body presents a certain amount of resistance to the flow of electric current. This however is not a constant value. It depends on factors such as body weight and the manner in which contact occurs and the parts of the body that are in contact with the earth. Figure 3.1 illustrates this point.

If the flow of current through the human body involves the heart muscles, it can produce a condition known as fibrillation of the heart denoting cardiac malfunction. If allowed to continue, this can cause death. The threshold of time for which a human body can withstand depends on the body weight and the current flowing through the body. An empirical relation has been developed to arrive at this value:

$$T_S = \frac{S_B}{I_B^2}$$

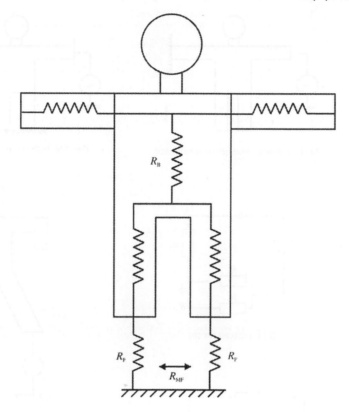

Figure 3.1
Resistance of human body to current flow

where T_S is the duration of exposure in seconds (limits of 0.3 and 3 s), I_B is the RMS magnitude of current through the body and S_B is the empirical constant.

Using this relation and assuming a normal body weight of 70 kg, it can be calculated that:

$$I_B = \frac{0.157}{\sqrt{T_S}}$$

where I_B is the RMS magnitude of current through the body (A) and T_S is the duration of exposure in seconds (decided by the operation of protective devices).

This value, however, has to be used with care. For example, a considerable portion of the body resistance is due to the outer skin. Any loss of skin due to burning in contact with electrical conductors can lower the resistance and increase the current flow to dangerous values.

In general, two modes of electrical potential application can happen. The first case is when a person is standing on the ground and touching an electrically live path. The other is the case of a potential difference between two points on the ground being applied across the 2 ft with the distance being about 1 m. Refer Figure 3.2, which illustrates these conditions.

Since the human body presents different values of resistance to the flow of electricity in these two modes, the voltage limits for tolerance of human body are calculated individually for both cases as follows.

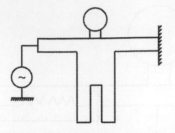

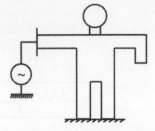

(a) Potential applied between two hands

(b) Potential between one hand and both feet

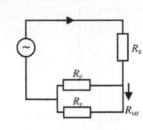

(c) Eq. circuit for (b)

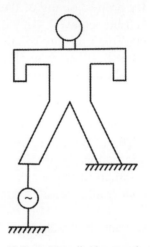

(d) Potential applied between the feet

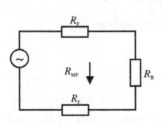

(e) Eq. circuit for (d)

R_B: Body resistance

R_F: Contact resistance of one foot and ground

R_{MF}: Mutual resistance between both feet

Figure 3.2
Modes of application of electric potential

Case 1   Contact with live part by hand

$$R_A = R_B + 0.5(R_F + R_{MF})$$

where R_A is the touch voltage circuit resistance (Ω), R_B is the body resistance (taken as 1000 Ω), R_F is self-resistance of each foot to remote earth in ohms and R_{MF} is the mutual resistance between the feet in ohms.

Case 2   Contact with feet

$$R_A = R_B + 2R_F - 2R_{MF}$$

where R_A is the step voltage circuit resistance in ohms, R_B is the body resistance taken as 1000 Ω, R_F is the self-resistance of each foot to remote earth in ohms and R_{MF} is the mutual resistance between feet in ohms.

The type of contact that normally happens in a building or other consumer installations is mostly of first mode. The voltage of tolerance in this mode as calculated in case 1 is called as touch potential. The occurrence of the second mode of contact is specific to outdoor electrical substations with structure mounted equipment and therefore is not much of relevance in our discussions. The voltage value arrived at for case 2 is known as step potential.

It therefore follows that the design of commercial, industrial and domestic electrical installations and their grounding methods should be done with due consideration to touch potential that can arise during abnormal or fault conditions.

3.3   Grounding of equipment

Electrical equipment grounding is primarily concerned with connecting conductive metallic enclosures of the equipment, which are not normally live to the ground system through conductors known as grounding conductors. For the grounding to be effective, the fault current (in the event of a failure of insulation of live parts within the equipment) should flow through the equipment enclosure to the ground return path without the enclosure voltage exceeding the touch potential. This is also applicable to other parts that are normally dead (refer Figure 3.3).

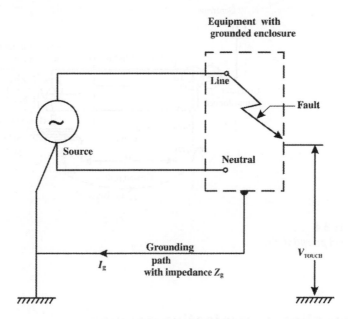

Figure 3.3
Voltage pattern during ground fault

The touch potential in such a case can be calculated by the application of Ohm's law:

$$V_{\text{touch}} = I_G \times Z_G$$

where I_G is the maximum ground fault current that is expected to flow and Z_G is the impedance of the ground return path.

I_g is usually determined by the type of system grounding adopted and the protective devices that are used for fault detection and isolation.

From the above, it will be clear that the impedance of the grounding conductor between the enclosure and the groundmass should be limited to a value as low as practically achievable in order to avoid dangerous potential levels appearing on the enclosures. This will ensure that accidental human contact with these enclosures will not result in fatal electrocution or serious injuries.

Another point to note is that in the case of a remote source without direct connection of metallic ground return path, the ground fault currents tend to flow through the groundmass. This causes an elevation of groundmass potential at the receiving end. Since the touch voltage is between enclosure and local groundmass, it is not of relevance as far as human safety is concerned (refer to Figure 3.4 for illustration). This condition is more relevant to three-wire systems in medium voltage systems where usually metallic ground return paths between source and receiving equipment are absent. In most low-voltage applications, this is not likely to be the case. In any case, the point to be remembered is that the potential rise of the enclosure with reference to local groundmass is what essentially matters to render the system safe, regardless of other issues involved.

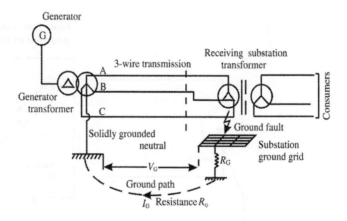

Figure 3.4
Ground potential rise

3.4   Operation of protective devices

When a fault to an enclosure takes place in electrical equipment, the return path through the groundmass alone is insufficient to operate the protective devices such as over-current release or fuses. This is so because the impedance between the enclosure and the groundmass is usually high enough to severely restrict the flow of fault currents, which is particularly true in low-voltage systems that are in common use. In these cases, it is imperative that a low-impedance ground return path to the source is available so that fault current of adequate magnitudes to cause operation of protective devices is ensured. The grounding conductor fulfills this function of a low-impedance connection. Figure 3.5 illustrates the point.

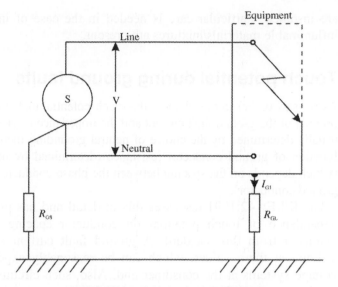

(a) Grounding path is completed through earth

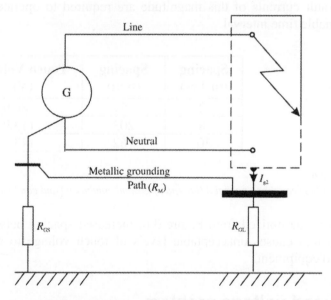

(b) Grounding path completed through metallic connection of very low impedance

Figure 3.5
Importance of ground return path

3.5   Thermal capability

It is essential that the grounding conductor mentioned in the previous paragraph or any other circuit component, which serves this function, should be designed to withstand the resulting fault current without developing excessive temperature or causing sparking at the joints. This will happen if the joints in the conductor or other bonding connections are improperly executed. This condition must be true for the magnitude and duration of the current required for protective devices to operate and isolate the faulty circuit.

It is therefore necessary to ensure good quality workmanship in these installations as otherwise the high temperatures or sparking may cause fires in the premises where they are installed. Particular care is needed in the case of installations where hazardous or inflammable materials/mixtures are present.

3.6   Touch potential during ground faults

As we saw earlier in this chapter, the touch potential that can develop on any enclosure is the product of the ground fault current and the impedance of grounding path. The current value is usually determined by the choice of neutral grounding method adopted. The impedance is a function of ground conductor resistance (determined by the conductor size) and reactance (which depends on the spacing between the phase conductor contributing to the fault and the ground conductor).

The IEEE:142:1991 discusses this in detail and has provided tables that illustrate the dependence of touch potential on conductor spacing. Figure 3.6 shows the details extracted from this standard. A ground fault current of 5 kA has been assumed for arriving at these values, which can be reasonably expected in solidly grounded low-voltage systems at the consumer end. Also, such circuits are not usually provided with sensitive ground fault protection and rely on over-current releases and fuses, in which case fault currents of this magnitude are required to operate protective devices within reasonable time interval.

Spacing (inches)	Spacing (mm)	Touch Voltage (V)
2	51	86.9
8	203	113.9
30	762	143.

Figure 3.6
Relation between touch potential (far end of ground conductor) and conductor spacing

As can be noticed from Figure 3.6, increased spacing between phases and grounding conductors causes unacceptable levels of touch voltage to appear on the enclosure of faulted equipment.

3.7   Induced voltage problem

The spacing between the phases and grounding conductor can also cause another problem. The magnetic flux, which is generated when ground fault current flows through the system can induce voltages in any nearby looped conductor with which it has a coupling. It is not necessary that this loop should be in physical contact with the electrical conductors. Mere magnetic linkage will be sufficient for this induction. Figure 3.7 shows this condition.

The voltage thus induced may not be very high but can drive a high current if this loop is closed by itself. In the above example, currents of 500 A and an open circuit voltage of 2.5 V can typically appear. The energy content of this induction will be enough to cause an explosion or fire if suitable explosive mixtures are present in the immediate environment.

In cases where high earth fault currents are of the order of 50 kA (typically outdoor substations of solidly grounded type), dangerous potential may appear across open ends of the loop and can pose an electrical hazard. This discussion illustrates the need for proper planning for running the grounding conductors so that touch potential and induced current hazards are eliminated.

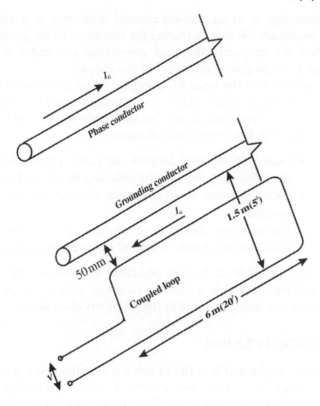

Figure 3.7
Example of induced potential

3.8 Mitigation by multiple ground connection

One way of mitigating the problem is by bonding the ground conductor at different points to the building structures. This will prevent the voltage becoming very high by providing multiple earth return paths. But this has the effect of transferring part of the potential to building structures. In other words, potential differences may be observed between different points of the building structures. Also, the partial flow of fault currents through joints in the structure that are not meant to conduct currents may cause heating or sparking during faults. This solution has therefore to be applied with adequate caution. For example, when reinforced concrete members are used for grounding purposes, it should be ensured that all reinforcing rods are properly connected together and joined with exothermic welds to the grounding earth conductor buried around the building. Proper attention to detail during the design and construction of high-rise buildings can result in very low grounding impedances on almost every floor in a tall building.

3.9 Mitigation by reduction of conductor spacing

Since the touch potential during faults is dependent on the spacing between ground return conductor and the phase conductor, it follows that a reduction of spacing will have the effect of reducing the touch potential. One way to do this is to run the grounding conductor bundled with the phase conductors. This will cause the reactance of the ground path to become very low.

Another way is to arrange the ground conductor as a metallic sleeve surrounding the phase conductor. Besides reducing the reactance of the ground path, this has the effect of canceling the magnetic flux and preventing any induced voltages from appearing in external loops as discussed earlier in this chapter.

This approach is the basis for using the metal raceway as the ground conductor. Use of high section rigid steel conduits as grounding conductors has the effect of reduction of touch potential and elimination of magnetically induced voltages in external circuits.

The following, however, needs attention:

- Excessive raceway lengths can cause problems by increasing the drop along the grounding path and by reducing the flow of fault currents.
- Raceway joints should be made without introducing any appreciable electrical resistance.
- The raceway should have adequate cross-sectional area for carrying ground fault currents for the length of time required for protection operation.

It is also possible to have a dedicated grounding conductor running along with the phase conductors within a metallic raceway, which also acts as a parallel ground return path which can mitigate some of the problems cited above.

3.10   EMI suppression

One of the unplanned benefits of using a metallic conduit as grounding conductor and as wiring raceway is that any electrical noise emanating from the electrical system conductors will get suppressed. The conduit acts like a screen for electromagnetic flux, which get trapped within the screen and does not radiate outside the enclosure.

This is of particular relevance in modern electrical systems where use of power semiconductors creates harmonic current flow through the system as well as line voltage notching, which can act as noise source by radiation from electrical power conductors. When the conductors supplying power to such equipment are enclosed within the metallic raceway, such electromagnetic interference (EMI) automatically gets suppressed.

3.11   Metal enclosures for grounding conductors

The earlier discussion was about the use of metallic raceway surrounding the phase conductors as grounding conductors for improved performance. But the use of a protective metallic sleeve around a grounding conductor from the service ground point to a grounding electrode presents a different problem.

These conductors carry current only when there is a ground fault and carry current one way. The other part of the current flows in a different circuit remote from the grounding conductor. Providing a steel protective sleeve, which is a magnetic material around this conductor, has the effect of increasing the reactance of the conductor by a factor of about 40.

Take, for example, a coil wound on a former without core connected to an AC supply. Now put a magnetic core within the former. You will notice that the current drops sharply because of the increased inductance. The pipe sleeve behaves in a similar fashion as the core (refer Figure 3.8).

To avoid this problem, it is necessary to bond the grounding conductor at both entrance and exit points with each integral section of the metallic enclosure. This results in reduction of impedance and therefore the voltage drops. Simultaneously, the metal sleeve also acts as a parallel grounding conductor and causes the voltage drop to reduce further (refer Figure 3.9).

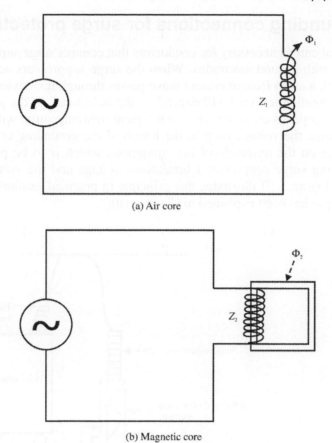

(a) Air core

(b) Magnetic core

Impedance of coil having magnetic core Z_2 is several times that of coil
with an air core Z_1 as $\Phi_2 > \Phi_1$ by several orders of magnitude.

Figure 3.8
Coil with and without a core

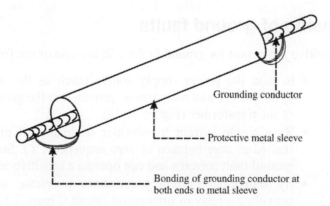

Grounding conductor

Protective metal sleeve

Bonding of grounding conductor at
both ends to metal sleeve

Figure 3.9
Bonding of earth conductor within a pipe sleeve

3.12   Grounding connections for surge protection equipment

Special care is necessary for conductors that connect surge suppression equipment ground leads with ground electrodes. When the surge suppressors act to conduct line surges to ground, a steep fronted current wave passes through the device to ground. The voltage of the grounding terminal will depend on the inductance of the grounding conductor, which in turn depends on its length. For a typical lightning surge with a rate of rise of typically 10 kA/µs, the voltage drop in the length of the grounding conductor is substantial. The voltage on the terminals of the equipment, which is to be protected, is the sum of the lightning surge suppressor's breakdown voltage and the voltage drop in the grounding wire. Figure 3.10 illustrates this principle (a practical example of the importance of this principle has been explained in Chapter 10).

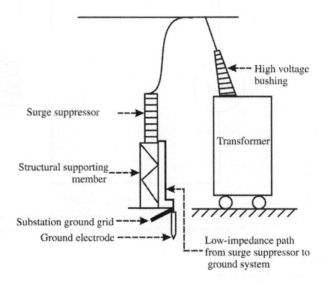

Figure 3.10
Ground connection of surge suppressor of a transformer

3.13   Sensing of ground faults

A sensitive protection for ground faults will use one of the following approaches:

- In case the power supply source (such as the transformer) is a part of the system, a CT and relay can be provided in the ground connection of the neutral of the transformer (Figure 3.11a).
- By a single current transformer enclosing all phase and neutral conductors (called as core balance or zero sequence CT). Such a transformer detects the ground fault currents and can operate a sensitive relay (Figure 3.11b).
- By individual current transformer in phase and neutral conductors and providing a relay in summation circuit (Figure 3.11c).
- Adding special ground fault equipment to the system to sense even low value of earth fault currents and trip the circuit faster.

Inclusion of the neutral in Figures 3.11b and c is for canceling any unbalance currents that may flow in the neutral from being sensed as ground faults.

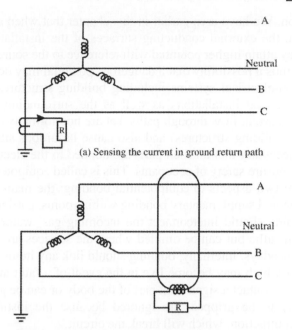

(a) Sensing the current in ground return path

(b) Zero sequence (core balance) transformer used to sense ground fault

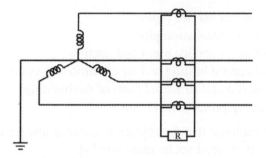

(c) Using summation of currents in three phases and neutral

Note: R is the ground fault sensing relay.

Figure 3.11
Sensitive ground fault protection connections

3.14 Equipotential bonding

Equipotential bonding is essentially an electrical connection maintaining various exposed conductive parts and extraneous conductive parts at substantially the same potential. An earthed equipotential zone is one within which exposed conductive parts and extraneous conductive parts are maintained at substantially the same potential by bonding, such as that, under fault conditions, the difference in potential between simultaneously accessible exposed and extraneous conductive parts will not cause electric shock.

Bonding is the practice of connecting all accessible metalwork – whether associated with the electrical installation (known as exposed metalwork) or not (extraneous metalwork) – to the system earth. In a building, there are typically a number of services other than electrical supply that employ metallic connections in their design. These include water piping, gas piping, HVAC ducting, and so on. A building may also contain steel structures in its construction. We have seen earlier in this chapter that when an earth fault takes place in an installation, the external conducting surfaces of the installation and the earth mass in the vicinity may attain higher potential with reference to the source earth.

There is thus a possibility that a dangerous potential may develop between the conducting parts of non-electrical systems including building structures and the external conducting parts of electrical installations as well as the surrounding earth. This may give rise to undesirable current flow through paths that are not normally designed to carry current (such as joints in building structures) and also cause hazardous situations of indirect shock. It is therefore necessary that all such parts are bonded to the electrical service earth point of the building to ensure safety of occupants. This is called equipotential bonding.

There are two aspects to equipotential bonding: the main bonding where services enter the building and supplementary bonding within rooms, particularly kitchens and bathrooms. Main bonding should interconnect the incoming gas, water and electricity service where these are metallic but can be omitted where the services are run in plastic, as is frequently the case nowadays. Internally, bonding should link any items, which are likely to be at earth potential or which may become live in the event of a fault and which are sufficiently large that they can contact a significant part of the body or can be gripped. Small parts, other than those likely to be gripped, are ignored because the instinctive reaction to a shock is muscular contraction, which will break the circuit.

In each electrical installation, main equipotential bonding conductors (earthing wires) are required to connect to the main earthing terminal for the installation of the following:

- Metal water service pipes
- Metal gas installation pipes
- Other metal service pipes and ducting
- Metal central heating and air-conditioning systems
- Exposed metal structural parts of the building
- Lightning protection systems.

It is important to note that the reference above is always to metal pipes. If the pipes are made of plastic, they need not be main bonded.

If the incoming pipes are made of plastic but the pipes within the electrical installation are made of metal, the main bonding must be carried out, the bonding being applied on the customer side of any meter, main stopcock or insulating insert and of course to the metal pipes of the installation.

Such bonding is also necessary between the earth conductors of electrical systems and those of separately derived computer power supply systems, communication, signal and data systems and lightning protection earthing of a building. Many equipment failures in sensitive computing and communication equipment are attributable to the insistence of the vendors to keep them separated from the electrical service earth. Besides equipment failures, such a practice also poses safety hazards particularly when lightning discharges take place in the vicinity. In such cases, large potential difference can arise for very short periods between metal parts of different services unless they are properly bonded. Some of the case studies in a later chapter deal with this issue.

If the incoming services are made of plastic and the piping within the building is of plastic, then no main bonding is required. If some of the services are of metal and some are plastic, then those that are of metal must be main bonded.

Supplementary or additional equipotential bonding (earthing) is required in locations of increased shock risk. In domestic premises, the locations identified as having this increased shock risk are rooms containing a bath or shower (bathrooms) and in the areas surrounding swimming pools.

There is no specific requirement to carry out supplementary bonding in domestic kitchens, washrooms and lavatories that do not have a bath or shower. That is not to say that supplementary bonding in a kitchen or washroom is wrong but it is not necessary.

For plastic pipe installation within a bathroom, the plastic pipes do not require supplementary bonding, and metal fittings attached to these plastic pipes also would not require supplementary bonding. However, electrical equipment still does require to be bonded and if an electric shower or radiant heater is fitted, they will require supplementary bonding as well.

Supplementary bonding is carried out to the earth terminal of equipment within the bathroom with exposed conductive part. A supplementary bond is not run back to the main earth. Metal window frames are not required to be supplementary bonded unless they are electrically connected to the metallic structure of the building. Metal baths supplied by metal pipes do not require supplementary bonding if all the pipes are bonded and there is no other connection of the bath to earth. All bonding connections must be accessible and labeled: SAFETY OF ELECTRICAL CONNECTION – DO NOT REMOVE.

3.15   Summary

In this chapter, we learnt about the basic issues of electric shock hazards. We reviewed the need for providing grounding connections to electrical equipment and the methods adopted for grounding. We discussed the ways and means of limiting the potential on equipment enclosures to safe values by proper ground connections. We also learnt how sensitive earth fault protection can be provided and where such protection is likely to be required. Information regarding grounding of substations and equipotential bonding of residential buildings was also covered. The salient provisions of various international codes relating to grounding and analysis of grounding behavior in MV/LV systems are discussed in Appendix A and they may be referred for additional information.

4

Lightning, its effect on buildings and electrical systems and protection against lightning

4.1   Introduction

Lightning is one of the most widely studied and documented natural phenomena. It is also one of the main causes of transient over-voltages in electrical systems. A proper understanding of lightning is essential for planning protection against lightning strikes so that no untoward damage is caused to buildings and electrical installations.

A lot of research has been done over a number of years worldwide and several publications as well as national and international standards have evolved which give us a good insight into this phenomenon.

Some of these are:

- AS 1768: 1991 Australian standard on lightning protection
- ANSI/NFPA 780 National lightning protection code
- IEEE 142: 1991 IEEE green book (Chapter 3)
- IEC 1024:1993 Protection of structures against lightning.

Lightning is the sudden draining of charge built up in low-cloud systems. It may involve another cloud system (which is not of much interest to us in this book) or ground (which is). The flow of charge creates a steep fronted current waveform lasting for several tens of microseconds. The flow is more usually that of negative charges though at times it may involve positive charge flow too. The latter are generally of lower magnitude and hence not critical to this discussion. Figure 4.1 shows a typical lightning waveform.

Some of the parameters of interest to us are:

- Peak current I usually expressed in kiloamperes
- Rate of rise of current dI/dT in kiloamperes per microsecond
- Time to crest T_{CR} in microseconds
- Time to fall to half of peak value T_H.

As the rate of rise is not uniform throughout, this value is further expressed as dI/dT (Max), dI/dT (10/90%) and dI/dT (30/90%).

dI/dT (Max) is the maximum value of slope in the rise curve, dI/dT (10/90%) is the average slope between 10% of peak current and 90% of peak current and so on.

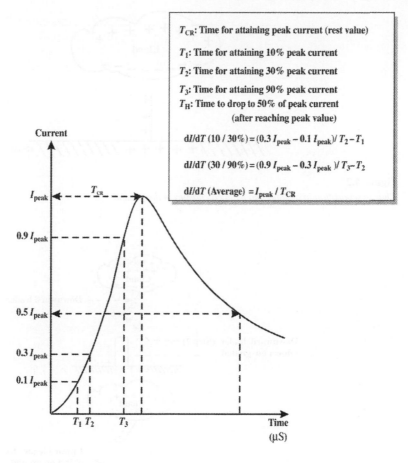

Figure 4.1
Typical waveform of lightning

The occurrence of lightning flash starts with a buildup of charge in a cloud system close to the ground. This charge is usually of the order of several million volts and usually of negative polarity at the bottom. Though the exact mechanism of charge separation is not clear, observations indicate that the ice particles in the top portion of the cloud are positively charged whereas the heavier water particles in the bottom portion of the cloud carry a negative charge.

Figure 4.2 shows the charge separation in a cloud and the corresponding induced charges in the ground. The movement of clouds causes corresponding movement of positive charges on the ground. This is observable as current flow in metallic pathways such as pipelines on the ground.

The high electrical field causes ionization of air and creates a conducting path. This usually happens near the cloud and the ionized path is called the downward leader. The leader precedes in steps of 20–30 m toward the ground, each step forming further ionization of the subsequent step. Simultaneously, from the high points or structures on the ground upward leaders of positive charges start forming. The interception of downward leaders and upward leaders completes the conducting path between the cloud and the ground and results in a lightning strike (see Figure 4.3).

The lightning flash along the ionized path causes a very high peak of current amounting to several kilo amperes and dissipates its energy in the form of heat (temperatures up to 20 000°C for a few microseconds), sound and electromagnetic waves (light, magnetic fields and radio waves).

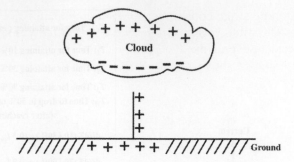

Figure 4.2
Charge separation in a cloud and induced charges in the ground

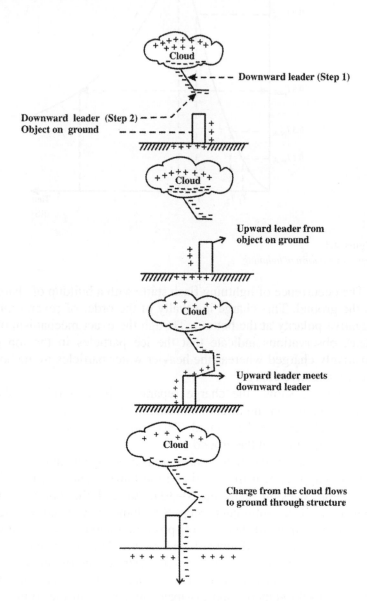

Figure 4.3
Lightning strike initiation

Each lightning strike consists of several component parts (each is a current wave called a stroke). The first current wave with a relatively lower dI/dT rate but higher in magnitude and several more (on an average 3) of much higher dI/dT but lower peak currents. Figure 4.4 shows a typical lightning discharge.

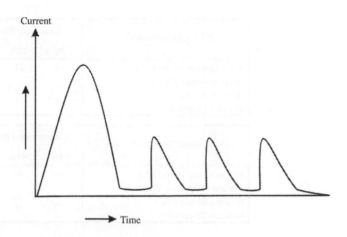

Figure 4.4
A typical lightning discharge with multiple wave components (strokes)

Like any typical natural phenomena, lightning strikes are not all identical but vary widely in their parameters. The values of these parameters are therefore defined in a probabilistic format. The tables shown in Figures 4.5 and 4.6 are examples of such variation. Figure 4.5 shows a table for peak lightning current. Figure 4.6 shows the maximum and average rate of rise of lightning current.

Type of Lightning Stroke	Cumulative Frequency				
	98%	95%	80%	50%	5%
First negative (kA)	4		20		90
Subsequent negative (kA)		4.6		12	30

Figure 4.5
Peak lightning current values

4.2   Incidence of lightning

Lightning depends on both atmospheric and geographical factors. It is usually associated with areas having convection rainfall. It requires presence of high moisture levels in air and high surface temperatures on ground. For example, the incidence of lightning is very high in Florida whereas in colder locations such as Canada where moisture levels in atmosphere are equally high are much less prone to lightning.

Since the protection to be given to buildings is a function of the probability of lightning strikes, the frequency of lightning occurrence has been extensively studied and the results are published in the form of annual isokeraunic maps for different world regions. These are contour maps, which show the mean annual thunderstorm days of the region involved. A thunderstorm day for this purpose is defined as one when thunder is heard at the point where it is measured. This obviously cannot indicate whether it is a result of inter-cloud or cloud to ground discharge. It does also show the frequency/number of instances or severity of cloud to ground strikes. Further studies are under way to gather such data and may result in modifications to the present methods of lightning risk assessment.

The isokeraunic map for Australia and New Zealand as well as Continental USA and Canada are shown in Figures 4.7a–e.

First Lightning Stroke	Cumulative Frequency		
	95%	50%	5%
Maximum rate of rise kA/μs	9.1	24	65
Average steepness kA/μs			
Between 30 and 90%	2.6	7.2	20
Between 10 and 90%	1.7	5	14

Subsequent Negative Strokes	Cumulative Frequency		
	95%	50%	5%
Maximum rate of rise kA/μs	10	40	162
Average steepness kA/μs			
Between 30 and 90%	4.1	20	99
Between 10 and 90%	3.3	15	72

Figure 4.6
Maximum and average rate of rise of lightning current

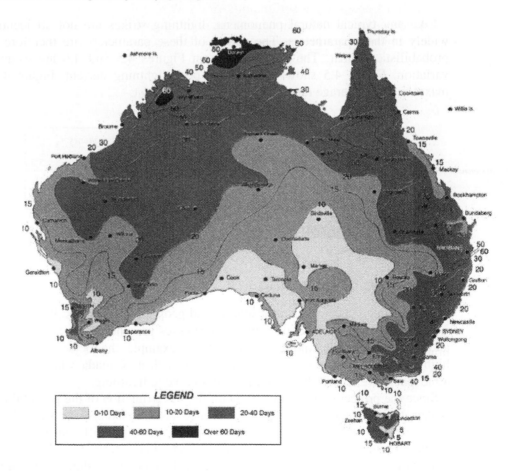

Figure 4.7a
Isokeraunic map of Australia

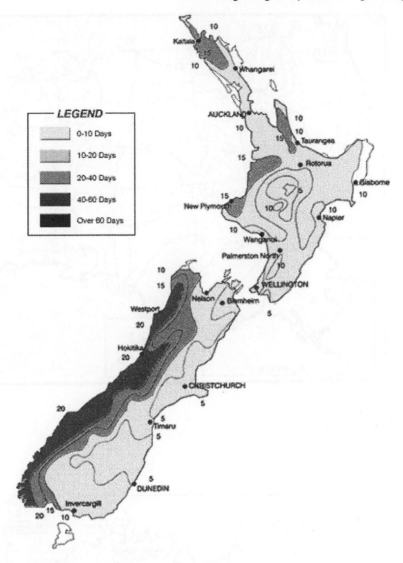

Figure 4.7b
Isokeraunic map of New Zealand

Figure 4.7c
Isokeraunic map of USA

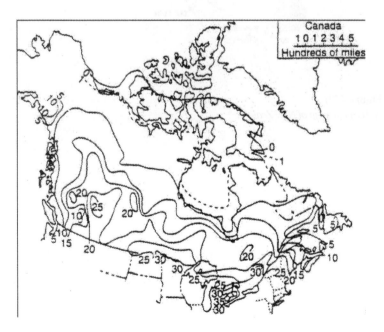

Figure 4.7d
Isokeraunic map of Canada

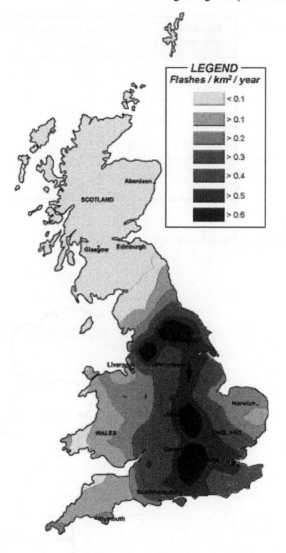

Figure 4.7e
Isokeraunic map of the UK

Methods have been developed to compute the ground flash density (flashes/ km²/year) from the average thunderstorm days. Table in Figure 4.8 gives the relationship between these parameters.

4.3   Probability of lightning strike

The probability of lightning strike depends on two factors: the incidence of lightning strikes in the geographical area where the structure is situated and the attractive area offered by the structure for lightning. The attractive area can be defined as the horizontal area within which a downward leader will be intercepted by an upward leader originating from the structure. Figure 4.9 shows the attractive area for a lightning mast. The attractive area in turn depends on the attractive radius R_A (shown in Figure 4.9). If the downward leader of a lightning comes anywhere within the sphere formed by the attractive radius with the top of the mast as center it will strike the mast.

Thunderstorms Days per Year	Ground Flash Density (in Flashes/km^2/year)	
	Average	Limits
5	0.2	0.1–0.5
10	0.5	0.2–1.0
20	1.1	0.3–3.0
30	1.9	0.6–5.0
40	2.8	0.8–8.0
50	3.7	1.2–10
60	4.7	1.8–12
80	6.9	3.0–17
100	9.3	4.0–20

Figure 4.8
Relationship between thunderstorm days and ground flash density

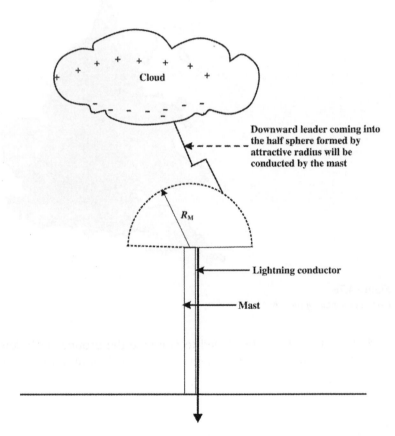

Figure 4.9
Attractive radius of a lightning mast

An empirical formula for the attractive radius is:

$$R_A = 0.84 \times h^{0.6} \times I^{0.74}$$

Where R_A is the attractive radius in meters, h is the height of the lighting mast in meters and I is the peak lightning current in kiloamperes.

For a horizontal conductor (such as the shield wire provided on an overhead electrical line) the attractive distance L_A is given by the formula:

$$L_A = 0.67 \times h^{0.6} \times I^{0.74}$$

Where L_A is the distance of attraction on either side of the conductor in meters, h is the height of the conductor from ground in meters and I is the peak current of lightning in kiloamperes.

The number of lightning strikes that a structure will attract in 1 year can be arrived by multiplying the ground flash density (in flashes/km^2/year) with the attractive area expressed in km^2.

4.4   Method of lightning protection

Lightning protection to any building or structure consists of providing a safe low-impedance conducting path for flow of lightning discharge currents to ground without allowing them to flow through the building structures. In practical terms, such protection consists of an air termination, down conductors and ground electrodes. A lightning mast independent of the structure but near enough to divert any lightning occurring in the vicinity is one example of protection. Figure 4.10 illustrates how a nearby mast protects a building. The protection offered may not be complete if the attractive radius of the building to lightning extends well beyond that of the mast. In such a case, it is possible that some of the strikes may hit the building rather than the mast.

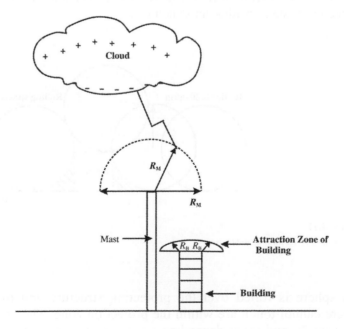

Figure 4.10
Building protected by a nearby mast

Whether a structure gets adequate protection from a lightning mast can be verified by using the principle of cone of protection. Figure 4.11 illustrates this principle. The angle A can vary between 30 and 60° depending on the degree of lightning protection desired for the structure (lower values for higher degree of protection).

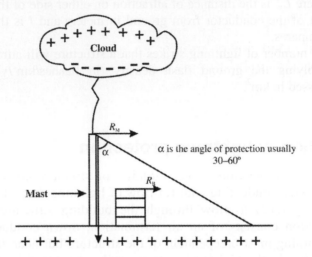

Figure 4.11
Example of the cone of protection principle

This concept is somewhat outdated and other better methods have been evolved to more accurately predict the protection offered. One such method is the rolling sphere principle of protection. Figure 4.12 illustrates the protection offered by a building's lightning protection system to adjacent structures.

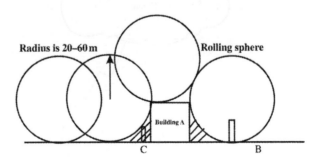

Figure 4.12
Rolling sphere method of protection

A sphere is rolled over the protecting structure and the shaded areas which the sphere cannot touch are within the protection zone. The radius of the sphere can vary between 20 and 60 m depending on the degree of protection required. The standard protection will consider a radius of 45 m and increased degree of protection can be obtained by reduction of the radius (refer Figure 4.13).

Protection Level	Radius of Sphere (m)
I	20
II	30
III	45
IV	60

Figure 4.13
Rolling sphere radius for different protection levels

Protection levels shown in Figure 4.13 are in descending order of importance of the structure to be covered (protection level I to be applied for structures requiring the greatest protection, and so on).

Since all buildings cannot be protected by a free standing mast, the more usual approach is to have the conductors placed on the building itself (called as air terminations) and connect them to the ground through down conductors. The air terminations can optionally be vertical spikes placed on the top periphery of the building. These air terminations can be connected by a flat conductor run on the roof thereby offering multiple paths for the lightning current to flow to ground. The down conductors are connected to dedicated ground electrodes to offer a short conducting path to ground. Figure 4.14 illustrates this arrangement.

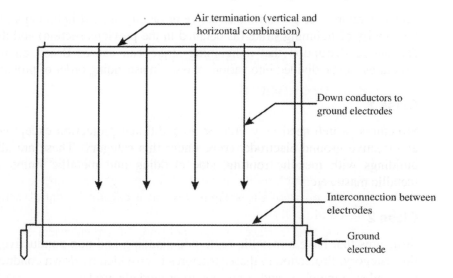

Figure 4.14
Typical lightning protection to a building

Lightning behaves like a current source. In other words, the flow of current is independent of the circuit impedance. If the current path to ground has high-impedance elements, the lightning current, which is of the order of several kiloamperes as we saw earlier, produces a very high voltage drop. This voltage appearing on the conducting elements can cause secondary flash to nearby earthed objects. It can also cause damage to building structures by forcing a path through non-conducting building elements. This explains why the lightning conductors should be of as low impedance or in other words as low a length as possible. Figure 4.15 illustrates the example of side flashes.

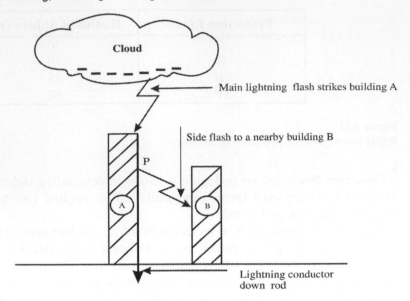

Figure 4.15
How side flashes are caused

4.5   Planning for lightning protection

The protection to be given for a structure or facility against lightning strikes is based on the probability of lightning strike (as detailed in the previous section) and the extent of risk of damage or disruption that a lightning strike can cause. Based on the latter criterion, structures can be divided into various classes in ascending order of protection requirement.

Class 1

Structures, which need very little or no additional protection except connecting them to an effective ground electrode, come under this category. These are all-metal structures, buildings with metallic roofing, side cladding and metallic frame work, stand-alone metallic masts, etc.

Class 2

Structures that have a metallic roof, side cladding and non-conductive framework are in this category. Protection to these structures is provided by down conductors bonded to the roof and side members and connected to ground electrodes.

Class 3

These include metallic frame buildings with non-metallic roof and side cladding. In this case, air terminations on the top of the building and on other non-conducting surfaces connected to the metal frame of the building are required to protect the insulating surfaces from being punctured by lightning.

Class 4

This class includes completely non-metallic structures such as buildings and tall chimneys/stacks constructed of reinforced concrete or masonry. These structures need extensive protection using air terminations, down conductors and grounding electrodes. An example of such protection is shown in Figure 4.16.

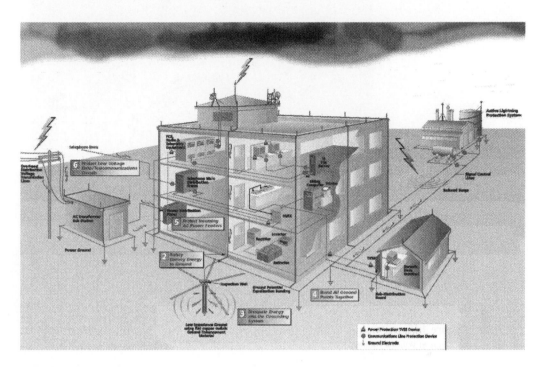

Figure 4.16
Example of lightning protection of a class 4 structure

Class 5

Buildings of historic or public importance or those containing valuable materials, places where a large number of people can gather at a time and public utilities such as power plants, water works, etc. come in this category and need utmost attention while planning protection.

4.6   Improvements to lightning protection

The protection to buildings and structures can be improved by better methods of prediction and by the use of active protection systems. We will cover them briefly below. In 1979, Eriksson presented an improved model, which allows for the intensification of ambient electric field created by a grounded structure. Eriksson's work was a fundamental step forward in lightning protection design, since it supported the field observations that the majority of lightning flashes terminate on the corners and nearby edges and other sharp features of unprotected structures, i.e. the points of highest electric field intensification. The ERICO scientists and engineers extended Eriksson's basic model for application to practical structures back in the late 1980s. This has been done through computer modeling of electric fields around a wide range of 3D structures and by application of the concept of 'competing features' to determine whether a structure is protected. This relatively new method has been known worldwide as the collection volume method (CVM) (refer to Figure 4.17).

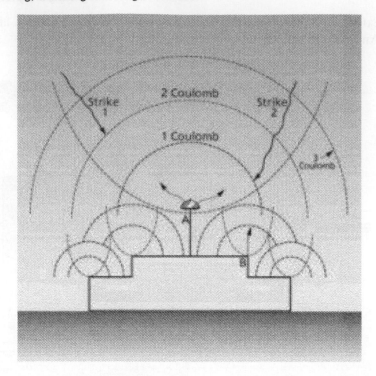

Figure 4.17
Collection volume method

The CVM takes the physical criteria for air breakdown, together with a knowledge of the electric field intensification created by different points on a structure and uses this information to provide the optimum lightning protection system for a structure, i.e. the most efficient protection design for the required protection level. Using the modern risk management approach, the CVM output depends on user-selected protection levels as per the previous rolling sphere method.

Active protection systems are also being offered by several vendors and are claimed to offer a higher degree of protection compared to the passive systems comprising air terminations and down-comers described earlier. The efficacy of many of these systems is however to be proven under actual installation conditions. The basic principle behind these systems is as follows. The active air terminations provided in these systems (which are vertical rods with an active component at their tip) generate a high electrical field as soon as a downward leader from a cloud starts toward the ground and immediately cause an upward leader to emanate from the air termination. Though a normal air termination also behaves in roughly the same fashion, the active protection systems react much faster. As a result the upward leader from the active air termination reaches out much higher resulting in the lightning strike to be invariably directed to the ground through the protection system.

4.7   Factors governing decision whether or not to protect

Standard AS 1768 provides clear guidelines to take a decision to provide or not to provide lightning protection to a building or structure based on an assessment of risk involved. The assessment is done in terms of the likelihood of the structure being struck and the consequences of any such strike. The use of the structure, the nature of its construction, the value of the contents and the prevalence of thunderstorms in the area can all be considered in making the assessment.

A decision to provide lightning protection may, however, be taken without any risk assessment, for example, where there is a desire that there should be no risk to a structure at all.

Examples of such structures are:

- Those in or near which large numbers of persons congregate
- Those concerned with the maintenance of essential public services
- Those in areas where lightning is prevalent
- Very tall or isolated structures and
- Structures of historic or cultural importance.

Where it is thought that the consequential effects will be small and that the effect of a lightning flash will most probably be merely slight damage to the structure, it may be economic not to incur the cost of protection but to accept the risk. Even then, it is better to make an assessment so as to give some idea of the magnitude of the risk that is being taken.

The need to protect electronic equipment and to protect persons against potential differences associated with metallic services increases with the building area. In such cases even though the construction of the structure does not warrant protection, appropriate measures must be taken to avoid risk to persons and equipment.

The standard also stipulates that any structure which is entirely within a zone protected by an adjacent object or objects (whether protected or not) should be deemed to be protected, that is no separate protection is necessary for such structures.

The standard defines a set of five indices:

1. Index A type of structure
2. Index B type of construction
3. Index C height of structure
4. Index D situation (location)
5. Index E lightning prevalence (thunderstorm days/year).

The sum of these indices (R) can be used to determine the need for protection. For more details, the relevant standard may be referred.

4.8   Effect of lightning strike on electrical lines

The foregoing discussion concentrated on the principles of lightning strikes and how their effects can be mitigated. However, lightning strikes on electrical lines or substations are those that cause problems in the distribution network which come right into our residences and offices. A full discussion on the protection of transmission and distribution lines from direct lightning strikes is beyond the scope of this book. We will, however, briefly touch upon this aspect further.

A direct strike on a conductor of a power line causes extremely high voltage pulses at the strike point, which are propagated as traveling waves in either direction from the point of strike. The crest of the pulse can be calculated as:

$$V = I \times Z$$

Where V is the crest voltage, I is the peak lightning current and Z is the impedance seen by the pulse along the direction of travel.

Impedance Z is equal to half the surge impedance of the line when struck at mid-point and can be approximately as much as $150\,\Omega$. Thus for a peak current of 40 kA, the voltage of the pulse can be as high as 6000 kV. Since the basic insulation level of most systems is much lower than this value, it is clear that such a pulse will cause failure of insulating components along the line. It is therefore necessary that no direct strike must be permitted on the overhead power line's phase conductors. This is achieved by stringing one or more shield wires along the phase conductors sufficiently above them so that the shield wires attract direct strikes and not the phase conductors. The shield wire is earthed at each transmission tower and thus the lightning current safely passes into the groundmass.

The clearance between the phase conductors and the shield wire must be selected so that air space between them does not breakdown by the high impulse voltage generated in the shield wires. This is easily achievable in systems of 66 kV and higher.

Even when protected in the above manner, the flow of the pulse of lightning current in the shield wire causes an induced voltage pulse in the phase conductors. These being much smaller in value than the direct pulse safely pass along the line without causing any insulation failure. To protect the equipment at the termination point of the overhead lines (such as circuit breakers, transformers, measuring devices, etc.), lightning arrestors are provided at the point of termination. These arrestors absorb any surges in the line and prevent them from traveling into the substation equipment.

These arrestors are essentially non-linear resistors in a porcelain housing which at normal voltages present a very high resistance. They are designed to break down at voltages above the highest system operating voltage (but lower than the basic insulation level of the system) thereby becoming good conductors and pass the energy of the lightning impulse to the ground. Once the voltage comes down (after the discharge of the pulse is over) the arrestors return to their original high-impedance state. The arrestors are placed on structures and their line terminals connected to each phase of the line. The other end of the arrestor (ground terminal) is connected to the substation grounding system through short ground conductors of adequate cross-sectional area.

Arrestors can also be optionally provided with surge counters for the purpose of monitoring their action.

4.9   Summary

In this chapter, we have reviewed the phenomenon of lightning, their effects on the installations in the ground. The probability of lightning strikes based on the ground geography and the configuration of the grounding installation was analyzed. We also went through the methods adopted for safeguarding the installations from the effects of a strike. The various classes of structures and how these are to be protected were also covered. The effect of lightning on electrical installations and the practices for ensuring their safety were also described.

5

Static electricity and protection

5.1 Introduction

Static electricity has been the earliest known form of electricity to mankind dating back to sixth century BC. Though in terms of power output, static electricity does not come anywhere near the electromagnetic form of electricity with which we are all familiar, it can be as much of a hazard as its other forms, and well-planned protection is needed against the hazards posed by it. In this chapter, we will discuss the physics of static electricity, its effects on equipment and humans and how to overcome the problems likely to be caused by it.

5.2 What is static electricity?

We know that all substances known to us are composed of molecules which are themselves formed from the atoms of a few basic elements and that all atoms are made up of subatomic particles, namely protons (positive charge), neutrons (neutral uncharged) and electrons (negative charge). In an atom, protons and electrons are equal in number thus making the atom electrically neutral and stable. Application of energy to a substance can cause separation of electrons from the parent atom. Electrons are free to move from the confines of their parent atoms whereas protons are bound to and move with the parent atoms and are therefore of limited mobility particularly so when the substance is a solid. Any substance which is deficient in electrons even marginally (1 in 100 000) exhibits a strong electric charge.

When two bodies of dissimilar materials are in contact with each other, electrons migrate from one body to the other through the contact surface. If the two bodies are suddenly separated, the electrons try to return back to their parent substance. In case the substances are electrically conductive, the electrons are able to do so, but if one of the substances or both are insulating materials this does not happen. The electrons get trapped in the surface of the material to which they have migrated. The surfaces of both substances now exhibit electrical charge due to excess or deficiency of electrons.

The voltage of a charged body can be calculated using the formula:

$$V = \frac{Q}{C}$$

where V is the voltage in volts, Q is the charge in coulombs and C is the capacitance of the body in Farads with reference to the surface of measurement.

The accumulated charge leaks away from the charged body gradually but if the rate of charge generation is higher than the rate of leakage, the voltage of the charged body increases till a breakdown of the surrounding medium takes place. This causes a spark to jump across the medium.

5.3   Generation of charge

Charge generation happens under various conditions. These are described below.

Type of materials

Static buildup needs two dissimilar materials, at least one of which should be an insulator, with differing dielectric constants to be in contact with each other.

Large contact area

The contact area between these dissimilar materials should be as high as possible to facilitate migration of electrons between the materials.

Speed of separation

The higher the speed of separation, less is the opportunity for the electrons to move back to the parent body and therefore higher is the charge buildup.

Motion between the substances

Though this is not a necessary condition, charge buildup is facilitated by motion between the surfaces in two ways. The friction and heat produced as a result of this friction increase the energy level of the atoms making escape of electrons easier. Secondly, the movement causes better surface contact by bringing the microscopic irregularities on both surfaces to come within contact of each other thus increasing migration.

Atmospheric condition

Humidity increases the leakage of charge through the air surrounding the charged body and reduces the buildup. Conversely the drier the atmosphere, the better is the ability to retain charge.

5.4   Some common examples of static buildup

A belted drive, especially of the flat variety on a pair of metallic pulleys rotating at high speed is one of the most common examples of charge buildup. The belt should be of non-conducting rubber material. Interestingly, conveyor belts made of rubber, which are used in material-handling systems extensively do not produce high static electricity due to the low linear speeds at which the conveyors work.

The speed of separation is not high enough to cause appreciable charge buildup.

- Flow of materials such as pulverized non-conducting solid particles, gas, air, etc. through a chute or an orifice results in charge generation.
- Rubber castors of chairs moving over non-conducting flooring and automobile tires running over road surface produce static charges.

- Motion of liquid falling into a tank during filling operations can produce static charges.
- Human workers can have several kilovolts of charge buildup on their bodies by the friction of rubber shoes or by charge-producing fabrics used in their dress.

The IEEE 142 (green book) gives the values of accumulation of charges for some of the commonly encountered conditions and the DC breakdown voltages and are reproduced in Figures 5.1 and 5.2 for comparison. It may be seen that the static voltage produced in these example situations is sufficient to cause a spark breakdown across gaps of up to 90 mm (3.5 in.). Also, it can be seen that a pointed object can cause spark at lower-voltage levels.

Type of Equipment/Process	Voltage Buildup (kV)
Belted drives	60–100
Fabric handling	15–50
Paper machines	5–100
Tanker trucks	<25
Belt conveyors handling grain	<45

Figure 5.1
Static voltage generation by various processes

Distance (mm)	Breakdown Voltage (kV) From a Point	Breakdown Voltage (kV) From a Plane
5	6	11
10	16	18
15	20	29
20	25	39
30	36	57
40	42	71
50	50	86
60	54	96
70	60	112
80	63	124
90	67	140

Figure 5.2
DC breakdown voltage

5.5 Energy of spark and its ignition capability

The energy of spark discharge can be calculated using the formula:

$$E = 0.5 \, (C \cdot V^2 \times 10^{-9})$$

where C is the capacitance of the body, which stores the charge in pF (picofarads), V is the voltage in volts and E is the energy in milliJoules.

The capacitances of some of the bodies discussed in the examples above are given below:

- Human body 100–400 pF
- Tank truck 1000 pF
- Tank with rubber lining 100 000 pF

The energy levels required to cause ignition depend on the flammability of the materials present in the environment and whether they form an explosive mixture. Too rich or too lean mixtures do not ignite easily.

5.6 Dangers of static electricity buildup

The following are the dangers posed by static electricity:

- Ignition causing fire or explosion
- Damage to sensitive electronic components
- Electric shock to humans followed by accidents such as a fall
- Damage to mechanical components such as bearings due to sparking through the oil films on bearing surfaces.

It is necessary to study the static buildup potential of any workplace and institute protective measures to control such buildup.

5.7 Control of static electricity

5.7.1 Grounding and bonding

As we have seen earlier in this chapter that charge buildup takes place when two surfaces, which are in contact and across which electrons migrate, get suddenly separated. Connecting such surfaces together with a conducting medium prevents charge accumulation by providing a leakage path. This is called bonding and can be achieved by using a bare or insulated conductor of adequate mechanical strength. Electric current flow due to charge leakage being of very low magnitudes, the size of the conductor is immaterial and so is the resistance of this conductor.

For moving objects, a ground brush of metal, brass or carbon can be used to provide the required leakage path. This method is commonly used for shafts of rotating machines to prevent bearing surface damages (refer Figure 5.3). For objects, which are in contact with ground already, no separate grounding or bonding is necessary.

Grounding cannot, however, provide a solution in all cases, especially where a bulky non-conducting material is involved. In this case, the part of the substance, which is a distance away from the grounded portion, can retain sufficient charge, since movement of charge will not be fast enough in an insulating material. This charge can result in a spark.

5.7.2 Control by humidity

Many insulating materials such as fabric, paper, etc. can absorb small quantities of water when the atmospheric humidity is sufficiently high. Even in the case of materials that do not absorb water, a thin layer of moisture gets deposited on the surface due to humidity (e.g., plate glass). If the environment has a humidity of over 50% moist insulating, materials can leak charges as fast as they are produced. This prevents high charge buildup thus avoiding sparks.

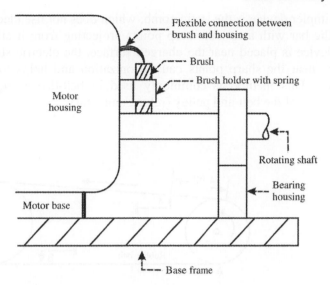

Figure 5.3
Example of the use of grounding brush

Conversely, most of the materials become dry when the humidity becomes lower than 30% since they tend to lose moisture to the atmosphere. This results in increased charge accumulation, which can cause sparking. Keeping humidity levels at 60–70% can solve static problems in many cases such as industries handling paper and fibers where charge buildup causes unwanted adhesion. In some cases, localized humidification using steam ejectors can be useful particularly where the large space involved makes increase of humidity in the entire space a difficult proposition.

This method is however unsuitable where:

- The processed material can be adversely affected due to high humidity.
- If the area involved is air conditioned or humidity controlled for process reasons or human comfort.
- In cases where humidity increase does not cause appreciable drop in resistivity.

In all such cases, other methods of static control may have to be resorted to.

5.7.3 Ionization

Ionization consists of forced separation of electrons from air molecules by application of electric stress or other forms of energy. The air thus ionized becomes conductive and can drain charges from charged bodies with which it is in contact. The positive ions and electrons are also attracted by the negative and positive charges respectively thus resulting in charge neutralization.

Ionization can be produced by high-voltage electricity, by ultraviolet light or by open flames. Various devices using a step up transformer operating on mains supply and producing high electric fields are commercially available. Due care is needed however to address safety issues arising out of the use of high voltage. Such devices find application in paper and fabric-processing plants. They are, however, unsuitable for use in situations where the environment contains inflammable gas mixtures.

A simpler device is the static comb, which does not use electricity at all. It consists of a metallic bar with a row of sharp points projecting from it and bonded to ground. When this device is placed near the charged surface, the electric stress due to accumulation of charge near the sharp points causes ionization and helps to drain the charge from the surface. This method is commonly used in belt-driven equipment near the point of separation of the belt and pulley (refer Figure 5.4).

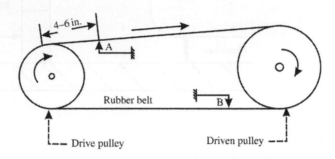

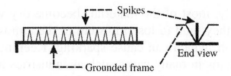

Figure 5.4
Use of static comb, an example

Another method of ionization is by using a row of small open flames. This method, however, requires caution where combustible materials are handled.

5.7.4   Use of anti-static (conductive) materials

Since the buildup of charge requires at least one of the surfaces to be non-conductive, it follows that by making the non-conductive surfaces conductive even slightly would reduce charge buildup. For example, coating a belt using conductive film on the side where it is in contact with the metallic pulley can prevent charge buildup.

Use of conductive flooring or conductive floor covering can reduce charge buildup. The resistance of the floor should be less than 1 MΩ when measured between points approximately 1 m apart for this method to be effective. At the same time, the resistance should be more than 25 000 Ω to avoid shock to personnel.

Similarly, conductive footwear and suits will prevent static buildup in the workplace. Also, materials with lower static producing properties should be used wherever necessary.

Static accumulation and discharge can destroy integrated circuit (IC) devices. Facilities, which handle these devices, or components that are made using them, should be designed with adequate precautions. Conducting cuffs connected with ground using metallic bonding conductors is a common device used in assembly shops to avoid transfer of charges from the operators' body to the circuit components.

5.8 Assessment of static risks and planning prevention

Workplace risks in respect of static charge buildup should be carefully assessed while planning a production facility.

The following questions need to be posed and answers sought.

- Is there equipment, which can cause generation of static charge in the planned facility?
- Does the workplace handle static-prone materials?
- Does the workplace involve processes where static generation is inherent?
- Are there flammable mixtures of gases or combustible materials present in the workplace?

Depending on the answers to the questions, the design of the facility should be reviewed and potential areas of static generation identified. Methods of static protection appropriate to the process in question should be selected and implemented. Wherever necessary, safe operating practices should be evolved, built into the working system and enforced. The equipment used for static prevention must be inspected from time to time, measurements made to verify their effectiveness and maintenance practices put in place to ensure their proper upkeep. A review should be carried out whenever process changes are effected.

These measures will greatly help in reducing the potential risks from static electricity.

5.9 Summary

In this chapter, we learnt about the physics of static electricity generation and the various factors that cause buildup of static charges. Conditions for spark generation from static voltage buildup were discussed. The methods of control of static buildup and preventing static spark discharge were also reviewed. The importance of making an assessment of static risk while planning a facility was also discussed.`1

6

Ground electrode system

6.1 Introduction

In the earlier chapters, we learnt about the need for connecting the power source neutral point to ground (system ground). We also discussed the requirement for a grounding connection at the consumer end, the equipment grounding. In both cases, the connection to ground or 'ground' requires the use of a system of ground electrodes. The effectiveness of grounding depends on obtaining as low a resistance as possible between the ground electrode system and the groundmass.

In this chapter, we will learn about the design of grounding system and the materials used for this purpose. The practices adopted in different countries follow the national standards/codes that are specified by the appropriate authority and can be significantly different. We will limit our discussion to the general principles involved in the design of earth electrode system. Specific examples of electrode system design are given in Appendix A for those who are interested. We will also discuss the methods of measuring the resistance of grounding systems.

6.2 Grounding electrodes

The construction of grounding electrodes depends on the local codes applicable. The purpose however is common. It is to establish a low-resistance (and preferably low-impedance) path to the soil mass. It can be done using conductors that are exclusively meant for this function or by structures/conductors used for other functions but which are essentially in contact with soil. However, while using the latter category, it must be ensured that the ground connection is not inadvertently lost during repair works or for any other reason.

6.2.1 Factors contributing to ground electrode resistance

The resistance of a ground electrode is made up of the following components:

- The resistance of electrode material
- Contact resistance of the electrode with soil
- Resistance of the soil itself.

The values of the first two are quite low compared to the last and can be neglected. We will discuss the third, namely resistance of the soil, in detail further.

6.3 Soil resistance

Though the ground itself being a very large body can act as an infinite sink for currents flowing into it and can be considered to be having very low resistance to current flow, the resistance of soil layers immediately adjacent to the electrode can be considerable.

Soil has a definite resistance determined by its resistivity that varies depending upon the type of soil, presence of moisture and conductive salts in the soil and the soil temperature. Soil resistivity can be defined as the resistance of a cube of soil of 1 m size measured between any two opposite faces. The unit in which it is usually expressed is ohmmeter.

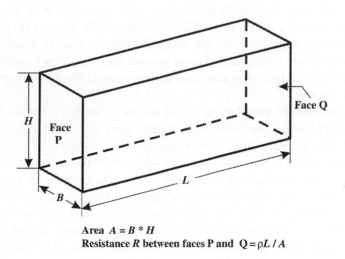

Area $A = B * H$
Resistance R between faces P and $Q = \rho L / A$

Figure 6.1
Soil resistivity

Resistance of the sample of soil shown in Figure 6.1 can be arrived at by the formula:

$$R = \rho L / A$$

Where

R: is the resistance between faces P and Q Ohm
A: Area pr faces P and Q (mz)
L: Length of soil sample in meter
ρ: Soil reisistivity Ohm-m

Soil resistivity for a given type of soil may vary widely depending on:

- The presence of conducting salts
- Moisture content
- Temperature
- Level of compaction.

Conducting salts may be present naturally in the soil or added externally for lowering the resistivity. Chlorides, nitrates and sulfates of sodium, potassium, magnesium or calcium are generally used as soil additives. However, the addition of such salts can be corrosive and in some cases undesirable from the environmental point of view. Especially, the presence of calcium sulfate in the soil is detrimental to concrete foundations and in case it is to be used for electrode quality enhancements, it should be limited to electrodes situated well away from such foundations. Also, over a period, they tend to leach away from the vicinity of the electrode. Moreover, these additive salts have to first get dissolved in the moisture present in the soil in order to lower the resistivity, and provision should be made for addition of water to the soil surrounding the electrode to accelerate this process particularly in dry locations.

Moisture is an essential requirement for good soil conductivity. Moisture content of the soil can vary with the season and it is advisable for this reason to locate the electrodes at a depth at which moisture will be present throughout the year so that soil resistivity does not vary too much during the annual weather cycle. It is also possible that moisture evaporates during ground faults of high magnitude for long duration. The electrode design must take care of this aspect. We will cover this in more detail later in this chapter.

Temperature also has an effect on soil resistivity but its effect is predominant at or near 0°C when the resistivity sharply goes up. Similarly, compaction condition of the soil affects resistivity. Loose soil is more resistive in comparison to compacted soil. Rocky soil is highly resistive and where rock is encountered, special care is to be taken. One of the methods of increasing soil conductivity is by surrounding the electrode with bentonite clay, which has the ability to retain water and also provides a layer of high conductivity. Unlike salts mentioned earlier, bentonite is a natural clay, which contains the mineral monmorillionite formed due to volcanic action. It is non-corrosive and does not leach away as the electrolyte is a part of the clay itself. It is also very stable. The low resistivity of bentonite is mainly a result of an electrolytic process between water and oxides of sodium, potassium and calcium present in this material. When water is added to bentonite, it swells up to 13 times of its initial volume and adheres to any surface it is in contact with. Also, when exposed to sunlight, it seals itself off and prevents drying of lower layers.

Any such enhancement measures must be periodically repeated to keep up the grounding electrode quality. A section later in this chapter describes about electrodes, which use these principles to dramatically lower the resistance of individual electrodes under extreme soil conditions. Such electrodes are commonly known as 'chemical electrodes'.

The IEEE 142 gives several useful tables, which enable us to determine the soil resistivity for commonly encountered soils under various conditions which can serve as a guideline for designers of grounding systems. These are shown in Figures 6.2 and 6.3.

6.4   Measurement of soil resistivity

Soil resistivity can be measured using a ground resistance tester or other similar instruments using Wenner's 4-pin method. The two outer pins are used to inject current into the ground (called current electrodes) and the potential developed as a result of this current flow is measured by the two inner pins (potential electrodes) (refer Figure 6.4).

Moisture Content (%)	Resistivity (in ohmmeter)		
	Top Soil	**Sandy Loam**	**Red Clay**
2	***	1850	***
4	***	600	***
6	1350	380	***
8	900	280	***
10	600	220	***
12	360	170	1800
14	250	140	550
16	200	120	200
18	150	100	140
20	120	90	100
22	100	80	90
24	100	70	80

Figure 6.2
Effect of moisture content on soil resistivity

Temperature (°C)	Resistivity (in ohmmeter)
–5	700
0	300
0	100
10	80
20	70
30	60
40	50
50	40

Figure 6.3
Effect of temperature on soil resistivity

The general requirements for ground resistance testing instruments are as follows:

- The instrument should be suitable for Wenner's 4-pin method. It should give a direct readout in ohms after processing the measured values of current injected into the soil and the voltage across the potential electrodes.
- The instrument should have its own power source with a hand-driven generator or voltage generated using batteries. The instrument will use an alternating current for measurement.

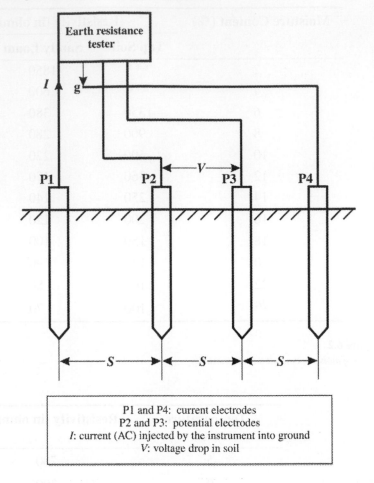

Figure 6.4
Soil resistivity measurement

- Direct reading LCD type of display is preferable. Resistance range should be between 0.01 and 1999 Ω with range selection facility for 20, 200 and 2000 Ω for better accuracy.
- Indications should preferably be available for warning against high current through probes, high resistance of potential probes, low source voltage and excessive noise in the soil.
- A minimum of four (4) steel test probes of length 0.5 m and sectional area of 140 mm^2 along with the necessary insulated leads (a pair of 30 m and another pair of 50 m) should be supplied with the instrument.

All the pins should be located in a straight line with equal separating distance between them and the pins driven to a depth of not more than 10% of this distance. Care should be taken to ensure that the connections between the pins and the instrument are done with insulated wires and that there is no damage in the insulation.

The resistance of the soil between potential electrodes is determined by Ohm's law ($R = V/I$) and is computed and displayed by the instrument directly. The resistivity of the soil is given by the formula:

$$\rho = 2\pi SR$$

Where

p: Soil resistivity in Ohm-m
S: Distance shown in Figure 6.2 in meters
R: Resistance measured in Ohms

Since the soil is usually not very homogeneous especially near the surface, the depth to which the pins are driven and the separation between the pins will cause resistivity figures to vary and can indicate the type of soil at different depths. The calculated value of resistivity can be taken to represent the value at the depth of 0.8*S* where *S* is the electrode spacing. The test is repeated at different values of *S* viz. 1, 2, 3, 5, 10 and 15 m and tabulated in the format shown in Table A.6 in Appendix A. They can also be plotted in the form of a graph.

A study of the values will give some indication of the type of soil involved. A rapid increase of resistivity at increasing *D* values shows layers of soil with higher resistivity. A very rapid increase may indicate the presence of rock and will possibly prevent use of vertical electrode. On the other hand, decrease of soil resistivity as D increases will indicate lower-resistivity soils in deeper layers where vertical electrodes can be installed with advantage.

In the case of any abnormality in the values, the test can be repeated after driving the pins along a different direction.

Errors can be caused by various factors in this measurement. They are as follows:

6.4.1   Errors due to stray currents

Stray currents in the soil may be the result of one or more of the following reasons:

- Differential salinity
- Differential aeration of the soil
- Bacteriological action
- Galvanic action (more on this later in the chapter)
- Ground return currents due to electric traction systems nearby
- Currents from multiple grounding of distribution system neutrals.

These stray currents appear as potential drop across the voltage electrodes without a corresponding current from the instrument's current source. Thus, they result in exaggerated resistivity measurements. This can be avoided by selecting an instrument source frequency, which is different from the stray currents, and providing filters that reject other frequencies.

6.4.2   Coupling between test leads

Improper insulation may give rise to leakage currents between the leads, which will result in errors. Ensuring good insulation and running the current and potential leads with a gap of at least 100 mm will prevent errors due to leakage.

6.4.3   Buried metallic objects

Buried metallic objects such as pipelines, fences, etc. may cause problems with readings. If presence of such objects is known, it will be advisable to orient the leads perpendicular to the buried object.

6.5   Resistance of a single rod electrode

The resistance of a ground electrode can be calculated once the soil resistivity is known. For a rod driven vertically into ground, the electrode resistance is given by the following formula

$$R = \rho / 2\pi L \{ \log \cdot (8L/d) - 1 \}$$

Where

R: is the resistance of the electrode in ohm
ρ: Soil resistivity in ohm meter
L: Length of electrode buried in soil
D: Outer dia of earth rod in m

A simplified formula for an electrode of 5/8 in. (16 mm) diameter driven 10 ft (3 m) into the ground is:

$$R = \rho / 3.35$$

Where

R: the resistance of the electrode in ohms
ρ: the soil resistivity in ohm meter

Knowledge of the soil resistivity alone is thus adequate to assess the electrode resistance to a reasonable degree of accuracy. The IEEE 142 gives the following table (Figure 6.5) for ready reference and can be used to arrive at the resistance value of the standard ground rod for different types of soil.

Soil Type	Average Resistivity (Ω m)	Resistance of Rod Dia. 5/8 in. Length 10 ft in ohms
Well-graded gravel	600–1000	180–300
Poorly graded gravel	1000–2500	300–750
Clayey gravel	200–400	60–120
Silty sand	100–800	30–150
Clayey sands	50–200	15–60
Silty or clayey sand with slight plasticity	30–80	9–24
Fine sandy soil	80–300	24–90
Gravelly clays	20–60	17–18
Inorganic clays of high plasticity	10–55	3–16

Figure 6.5
Soil resistivity for different soil types

6.5.1   Resistance distribution in soil surrounding a single electrode

The resistance of the soil layers immediately in the vicinity of the soil is significant in deciding the electrode resistance. To illustrate this let us see Figure 6.6.

A current that flows into the ground from a buried electrode flows radially outward from the electrode. It is therefore reasonable to assume for the purpose of calculating the soil resistance that the soil is arranged as concentric shells of identical thickness with the electrode at the center. The total resistance can thus be taken as the sum of the resistance of each shell taken in tandem.

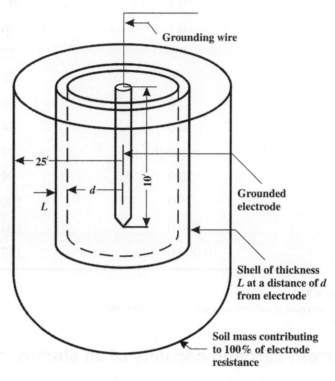

Figure 6.6
Soil resistance distribution around a vertically driven electrode

The resistance of each shell is given by the formula:

$$R = \rho L / A$$

Where

    R: Resistance of the shell
    L: Resistance of the shell
    A: Surface are (inner) of the shell

The area of the shells keeps increasing as we move away from the electrode. Thus, the resistance of the shells keeps reducing in value. The IEEE 142 has tabulated this variation (see Figure 6.7).

It can be seen from Figure 6.7 that the first 0.1 ft accounts for 25% of the resistance value and the first 1 ft for 68%. At 10 ft (equal to the rod length), 94% of the resistance value has been achieved. For this reason, lowering of soil resistivity in the immediate vicinity of the electrode is the key to lowering the electrode resistance. Also, placing more ground electrodes in the vicinity will only interfere with the conduction of current since the current from one electrode will increase the ground potential, which will have the effect of decreasing the current flow from the other nearby electrode (and vice versa).

Distance from Electrode (in feet)	App. % of Total Resistance
0.1	25
0.2	38
0.3	46
0.5	52
1.0	68
5.0	86
10.0	94
15.0	97
20.0	99
25.0	100
100.0	104
10 000.0	117

Figure 6.7
Radial variation of soil resistance around a rod electrode

6.6   Current-carrying capacity of an electrode

When current flow through a ground electrode into ground is low, the heat generated in the ground layers gets dissipated fairly fast and does not lead to any appreciable temperature rise. On the other hand, for a high current flow as happens during faults in solidly grounded systems, the effect would be quite different. As we saw earlier, the bulk of the resistance is concentrated in the immediate vicinity of the electrode. Without adequate time for the heat generated to be conducted away, the temperature of the ground layers surrounding the ground electrode rises sharply and causes evaporation of soil moisture around the electrode. If this persists, the soil around can become dry losing all the moisture present in it resulting in arcing in the ground around the electrode. Thus, a smoking or steaming electrode results in an electrode that is ineffective. To prevent this from happening, it is essential to limit the flow of current flowing into the ground through an electrode as indicated by the following formula:

$$I = \frac{(34\,800 \times d \times L)}{\sqrt{\rho \times t}}$$

Where

I: Maximum permissible current in amp
d: Diameter of the rod (m)
L: Buried length of rod (meters)
ρ: Soil resistivity ohm-m
t: Time of fault flow in second

6.7   Use of multiple ground rods in parallel

When it is not possible to obtain the minimum resistance stipulations or the ground fault current cannot be dissipated to the soil with a single electrode, use of multiple ground rods in parallel configuration can be resorted to. The rods are generally arranged in a straight line or in the form of a hollow rectangle or circle with the separation between the rods not lower than the length of one rod. As we have seen earlier in this chapter, the soil layers immediately surrounding the electrode contribute substantially to the electrode resistance. More than 98% of the resistance is due to a soil cylinder hemisphere of 1.1 times the electrode length. This is called the 'critical cylinder'. Placing electrodes close to each other thus interferes with the conduction of current from each electrode and lowers the effectiveness.

It is also of interest to note that the combined ground resistance of multiple rods does not bear a direct relationship to the number of rods. Instead, it is determined by the formula:

$$R = R/N \times F$$

R: Combined ground resistance of the electrode system having *N* electrode (Ohms)
R: Resistance of a single typical electrode (Ohms)
F: Factor *F* in the table shown in Figure 6.6 for Number of Roads = *N*

The table in Figure 6.8 shows the value of the factor *F* used above.

No. of Rods	F
2	1.16
3	1.29
4	1.36
8	1.68
12	1.80
16	1.92
20	2.00
24	2.16

Figure 6.8
Factor F *for multiple ground rods*

6.8   Measurement of ground resistance of an electrode

The resistance of a single ground electrode (as well as small grounding systems using multiple rods) can be measured using the 3-point (or 3-pin) method. The apparatus for this purpose is the same that is used for soil resistivity, viz. the ground resistance tester (see Figure 6.9). This method, however, may not yield correct results when applied to large grounding systems of very low resistance.

The measurement of electrode resistance is done in order to:

- Check on correctness of calculations and assumptions made
- Verify the adequacy after installation and
- Detect changes in an existing installation.

In this case, the ground electrode itself serves both as a current and potential electrode. The other electrode farther from the electrode is the other current electrode and the nearer one is the second potential electrode. The resistance can directly be read off the instrument. To get correct results, the current electrode must be placed at a distance of at least 10 times the length of the electrode being measured and the potential electrode at half the distance. A very similar method can be adopted for the measurement of ground grids, which are used commonly in HV substations (usually outdoor switchyards) (refer Figure 6.10).

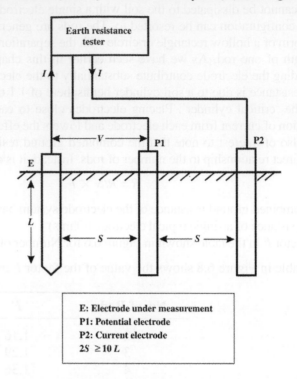

E: Electrode under measurement
P1: Potential electrode
P2: Current electrode
$2S \geq 10L$

Figure 6.9
Measurement of electrode resistance by 3-point method

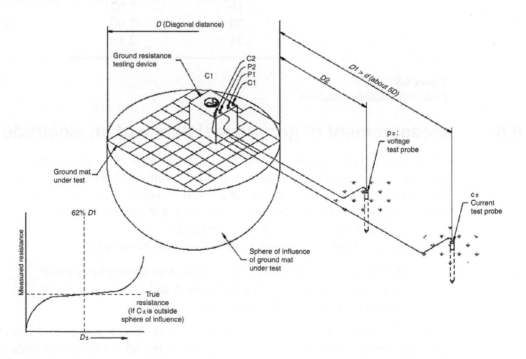

Figure 6.10
Measurement of resistance of a ground grid

The problems mentioned in the section on measurement of soil resistivity are applicable in this case too and appropriate precautions must be taken to ensure accuracy. A more detailed approach given in the South African standard SCSASAAL9 is described in Appendix A and can be used for better results.

6.9   Concrete-encased electrodes

Concrete foundations below ground level provide an excellent means of obtaining a low-resistance electrode system. Since concrete has a resistivity of about 30 Ω m at 20 °C, a rod embedded within a concrete encasement gives a very low electrode resistance compared to most rods buried in ground directly. Since buildings are usually constructed using steel-reinforced concrete, it is possible to use the reinforcement rod as the conductor of the electrode by ensuring that an electrical connection can be established with the main rebar of each foundation. The size of the rebar as well as the bonding between the bars of different concrete members must be done so as to ensure that ground fault currents can be handled without excessive heating. Such heating may cause weakening and eventual failure of the concrete member itself. Alternatively, copper rods embedded within concrete can also be used.

Concrete electrodes are often referred to as 'Ufer' electrodes in honor of Mr Ufer, who performed a large amount of research into concrete-encased electrodes. The rebars used are required to be either bare or zinc coated. Normally, the following applies to a rebar used as an earthing electrode:

- Minimum length of 6 m
- Minimum diameter 13 mm.

and installed:

- In a minimum of 50 mm of concrete
- Concrete is in direct contact with earth
- Located within and near the bottom of a concrete foundation or footing
- Permitted to be bonded together by the use of steel tie wire.

With respect to the last point, steel tie wire is not the best means of ensuring that the rebars make good continuity. Excellent joining products are available in the market, which are especially designed for joining construction rebars throughout the construction. By proper joining of the rebars in multi-level buildings, exceptionally good performance can be achieved. An extremely low-resistance path to earth for lightning and earth fault currents is ensured as the mass of the building keeps the foundation in good contact with the soil. Some examples of splicing products available in the market for jointing of rebars are shown in Figures 6.11a–c.

A recent advancement for solving difficult earthing problems is the use of conductive concrete to form a good earth. Normally, this form of concrete is a special blend of carbon and cement that is spread around electrodes of copper.

These are normally installed in a horizontal configuration by digging a trench of approximately a half a meter wide and 600 mm deep. The flat copper or rods are then installed in the center of the trench. The conductive concrete is then applied dry to the copper and spread to approximately 4 cm thick over the copper to the edges of the trench. The trench is then backfilled and the conductive concrete then absorbs moisture from the soil and sets to about 15 Mpa.

It is also possible to install these electrodes vertically. However, in this case, the conductive concrete has to be made up as a slurry and pumped to the bottom of the hole to displace water or mud.

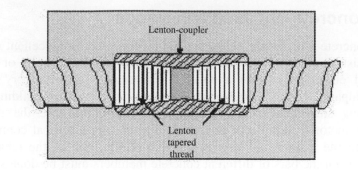

Figure 6.11a
Threaded splice joint using a coupler

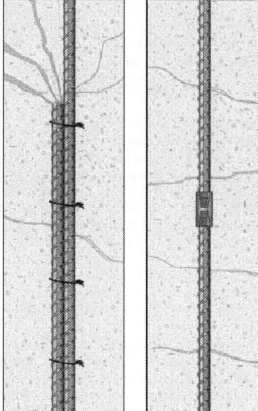

Lap splices depend on concrete for stregth and therefore lack structural integrity ana continuity in concrete construction.

Mechanical splicing provides the assurance of maintaining load path continuity of the structural reinforcement, independent of the condition or existence of the concrete.

Figure 6.11b
A comparison between lap splice using tie wire and threaded (mechanical) splice

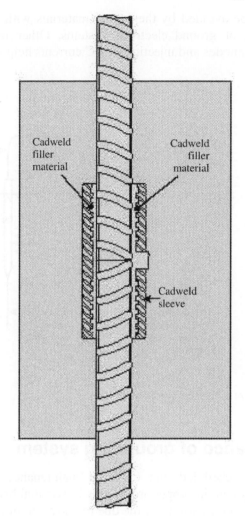

Figure 6.11c
A welded splice joint with sleeve

6.10   Corrosion problems in electrical grounding systems

Buried electrode systems bonded to other facilities embedded in ground such as piping/conduits can form galvanic cells when they involve dissimilar metals having differing galvanic potential. These cells, which are formed from the dissimilar metals as electrodes and the ground as the electrolyte, set up galvanic current through the bonding connections (refer Figure 6.12).

For example, copper electrodes and steel pipes used as a part of grounding system can cause cells of 0.38-V potential difference with copper as the positive electrode. This circulates a current, which causes corrosion of the metal in the electrode from which current flows into the ground. A galvanic current of 1-A DC flowing for a 1-year period can corrode away about 10 kg of steel.

This can be avoided by the use of materials with the same galvanic potential in the construction of ground electrode systems. Other methods such as use of sacrificial materials as anodes and injection of DC currents help to control this type of corrosion.

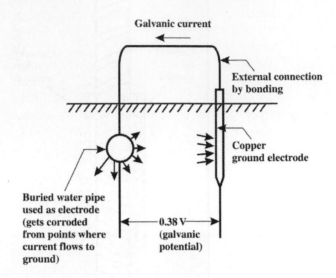

Figure 6.12
Galvanic action of a ground electrode system

6.11   Maintenance of grounding system

A properly scheduled and executed maintenance plan is necessary to maintain a grounding system in proper order. This is essential because the efficacy of the system can be affected over a period of time due to corrosion of metallic electrodes and connections. Periodic measurement of the ground electrode resistance and recording them for comparison and analysis later is a must. In the case of any problems, repairs or soil treatment must be taken up to bring the ground electrode system resistance back to permissible values.

6.12   Chemical electrodes

We have seen earlier in this chapter that the resistance of the ground electrode is influenced by the soil immediately surrounding the electrode. It is also influenced by the ambient conditions of the soil such as moisture and temperature. Thus, it is difficult to obtain acceptable values of grounding resistance in areas where:

- Natural soil is of very high resistivity such as rocky material, sand without vegetation, etc.
- During part of the year, the resistance may become excessive because of the absence of moisture.
- Soil temperature remains extremely low as in the case of polar regions or those close to the polar circle (called as permafrost condition, where the ground is below freezing temperature).

It thus follows that the performance of an electrode can be improved by using chemically treated soil to lower the soil resistivity and to control the ambient factors. While the soil temperature cannot be controlled, it is possible to ensure presence of moisture around the electrode. Soil treatment by addition of hygroscopic materials and by mechanisms to add water to the soil around the electrode are common methods of achieving this objective. Also, the resistivity behavior in permafrost conditions can be improved by soil conditioning, thus improving the electrode resistance dramatically.

Tests performed by the US Corps of Engineers in Alaska have proved that the resistance of a simple conventional electrode can be lowered by a factor of over twenty (i.e. 1/20). The treatment involved simply replaces some of the soil in close proximity with the electrode by conditioned backfill material. Refer Figure 6.13 for the result of tests conducted at Point Barrow, Alaska, which illustrates that the electrode resistance has dropped from a high of 20 000 Ω to a maximum of 1000 Ω by soil treatment.

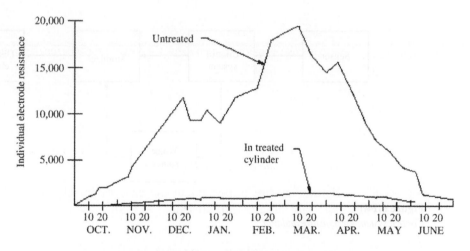

Resistance of electrodes, Oct. 1956–May 1957

Figure 6.13
Result of soil treatment on electrode resistance

The principle of improving the soil conductivity has been applied for a long time in ground electrode construction. The example shown earlier in Figure 1.5 belongs to this category. In this example, the hollow earthing tube contains sodium chloride, which absorbs moisture from surrounding air and leaches out to the soil to lower its resistivity. The backfill is soil mixed with charcoal and also sodium chloride. Since moisture in air is essential for this construction, means are provided to externally add water during dry weather.

These basic principles are used by several vendors who manufacture electrodes for applications involving problem areas. In these cases, both the electrode fill material and the augmented backfill are decided based on the soil properties so that moisture can be absorbed from surrounding soil itself and preserved in the portion immediately surrounding the electrode. In some systems, automated moisture addition devices are provided to augment this effect. A typical system by a vendor incorporating a solar-powered moisture control mechanism is shown in Figures 6.14a and b.

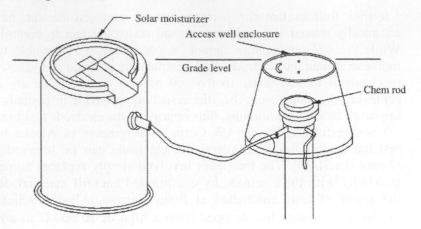

Figure 6.14a
Arrangement of chemical electrode with moisturizing mechanism

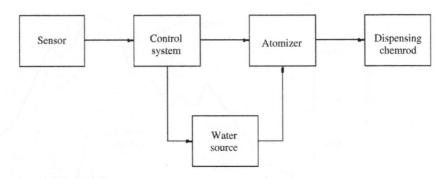

Figure 6.14b
Control system for moisture addition

6.13   Summary

In this chapter, we discussed the grounding electrodes that are commonly used in electrical systems and about their resistance. The details of soil resistivity and its measurement were examined. The methods to estimate the grounding resistance of single and multiple electrodes were reviewed. The measurement of ground electrode resistance was also covered. The need for limiting the ground current flow to prevent smoking electrode phenomenon was reviewed.

The improvement of soil to obtain better conductivity by use of conducting salts and bentonite clay was covered. Corrosion of ground electrode systems and the ways of preventing the same were covered briefly. Various tables, which will be of assistance in designing grounding systems, have been given in this chapter. These along with any applicable local codes may be deployed to ensure that the grounding system will function correctly and fulfill its primary purpose of achieving human safety. Appendix A at the end of this book gives an example of standardized ground electrode systems recommended by some national standards and procedure for measurement of electrode resistance.

7

Surge protection of electronic equipment

7.1   Introduction

In the previous chapters, we learnt about the necessity of grounding electrical equipment and non-current-carrying parts of electrical equipment for human safety under insulation failure situations. We also covered the basic facts about lightning, the need for lightning conductors and their grounding to protect a building and its inhabitants from lightning strokes and the accompanying surges during the discharge of lightning into ground. In this chapter, we will discuss the subject of surge protection in detail.

7.2   What is a surge?

A surge is a temporary but steep rise of voltage in a power system, usually as a result of lightning activity (but also sometimes due to internal causes). We had seen an example of such causes in an earlier chapter where the opening of current through an inductor was seen to give rise to a switching surge. The rise may last for a fraction of the power cycle wave. It may consist of a single spike or multiple diminishing spikes as we saw in Chapter 4 on lightning protection. A surge, unless properly protected against, causes failure of insulation in electrical wiring or devices due to excessive voltage. The energy contained in a surge may destroy the parts of a power system through which it passes (a result of the high magnitude of the current wave). Circuits with electronic components are especially highly vulnerable since the devices used in these circuits do not have much ability to withstand high voltages or currents.

Ways of protecting against surges are:

- Lightning protection systems as per relevant codes
- Quality grounding/earthing grid/bonding
- Surge arrestors on power circuits
- Multi-level protection on signal paths based on protection zone concept
- Continuous maintenance of systems.

7.3 Bonding of different ground systems as a means of surge proofing

In Chapter 3, we covered the basic principles of grounding the enclosures of electrical equipment to ensure safety of personnel in the event of ground faults. In Chapter 4, we covered in detail the physics related to lightning phenomenon and how the surges due to lightning strokes are safely conducted to ground using a lightning protection system consisting of air terminations, down conductors and grounding electrodes. Both these grounding systems are inherently noise prone, since conduction of surges and fault currents into ground is accompanied by a rise of voltage of the conducting parts connected to these systems with reference to the local earth mass. When sensitive electronic equipment first started appearing in the work place, it was usual for manufacturers of these equipment to demand (and get) a separate isolated ground reference electrode since it was claimed that connecting these systems with the building ground would affect their operation due to the ground noise. Thus, the concept of 'clean' ground was born as opposed to the other 'dirty' ground.

While this did give a solution of sorts to the problem of noise, it violated the fundamental requirement of personnel safety.

In Figure 7.1, we see three different types of ground each isolated from the others: the power system ground, the lightning protection ground and the 'clean' electronic ground. While this is perfectly trouble-free most of the time (when no lightning discharge or power system faults occur), the situation becomes positively dangerous when there is a surge due to lightning or faults. As we saw earlier, when lightning strikes a building, it produces a momentary high voltage in the grounding conductors due to the inherently fast rise time of the discharge and the inductance of the grounding leads (which is solely a function of their length). Similarly, when there is an insulation failure, the flow of substantial earth fault current causes a perceptible rise of voltage in the metal parts exposed to these faults and the associated grounding conductors (limited to safe touch potential values, but a rise all the same).

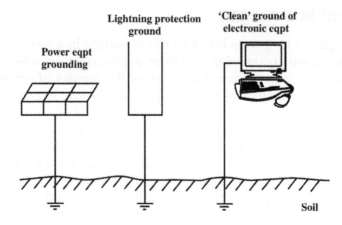

Figure 7.1
Isolated grounding systems

So while the clean ground which does not develop these high potentials remains at true ground potential, other metal parts or building structures or flooring in its vicinity can all assume a high potential, albeit briefly, during surges and faults. It means that a high potential can and does develop between the electronic ground and the equipment connected to it and the building ground or the lightning protection ground, which gives rise to inherently unsafe situations both for personnel and for the equipment connected to the 'clean ground'.

Another problem with an isolated ground is that the ground resistance of a system, which uses one or two electrodes, is much higher than the common ground. The touch potential of the electronic equipment enclosures in the event of an earth fault within the equipment may therefore exceed safe limits. The answer to these problems therefore lies in connecting all these different grounding systems together (refer Figure 7.2).

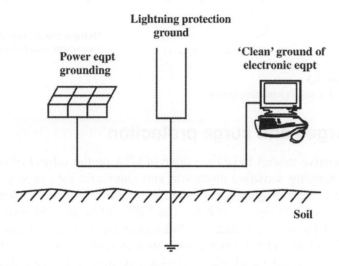

Figure 7.2
Grounding systems brought to a common electrode

Figure 7.2 shows all three grounding systems tied at a single point to the ground. Theoretically, this arrangement will prevent differential potentials between different grounds. But in practice, such a common ground electrode will have a high value of impedance, which cannot properly disperse lightning surges and will cause an undue potential rise in the grounding system with respect to the earth mass. The arrangement is therefore not of much practical value.

Figure 7.3 shows a system with multiple grounding points with different types of electrodes bonded together to form a low-impedance ground path which ties together all forms of grounding within the building. It prevents the grounding system from attaining dangerous potential rise with reference to the general earth mass and also avoids differential voltages between the building's exposed metallic surfaces and equipment enclosures.

It is this type of system that is installed in any modern facility to ensure that no unsafe conditions develop during lightning strokes or ground faults. The grounding system safely conducts away the surge currents led through by different surge protection devices into the ground path when surges occur in a system.

Bonding of the different earthing systems is thus the first step toward protection of sensitive equipment against surges. Practical examples of problems that can arise if this is not done are illustrated in Chapter 10.

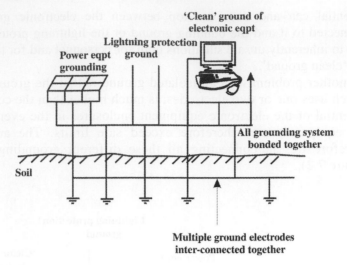

Figure 7.3
An integrated grounding system

7.4 Surges and surge protection

Extensive studies have been done in USA on the subject of surge protection requirements for housing sensitive electronic and automatic data processing (ADP) systems and the results were originally published in the form of a publication called Federal Information Processing Standards (FIPS). The FIPS publication 94 gave the guidelines on electrical power for ADP installations. Subsequently, these have been incorporated in the standard IEEE 1100 and FIPS publication 94 was withdrawn. The IEEE 1100 gives definitions of surge protective and related devices. A surge-protective device is defined as: A device that is intended to limit transient over-voltages and divert surge currents. It contains at least one non-linear component.

A surge suppressor is defined as: A device operated in conformance with the rate of change of current, voltage, power, etc., to prevent rise of such quantity above a pre-determined value.

A transient voltage surge suppressor (TVSS) is defined as: A device that functions as surge-protective device (SPD) or surge suppressor.

Any power system operates normally between certain voltage and frequency limits. Usual limits are ±10% of voltage and ±3% for frequency. These limits are valid only under 'normal' conditions. Abnormalities in the power system such as loss of major generation capacity, outage of a transmission line or a power transformer failure can cause these limits to exceed. A brownout is one such condition when the voltage becomes low on a sustained basis (called as 'sag'). A voltage 'swell' is the opposite of this condition with a voltage rise for prolonged duration.

In the event of system faults such as a short circuit or a ground fault, the system voltage can become much lower but for brief duration within which time, the protection systems come into operation and safely isolate the faulty circuit or equipment. In lower-voltage circuits, this is done by a fuse or miniature circuit breaker whereas larger power systems are provided with relays, which are extremely complex in nature. Once the isolation is complete, the system bounces back to normalcy very quickly. Between the appearance of a fault and its isolation, the voltage can dip as low as 10% of its normal value for a fraction of a second for close faults. For faults which are far away, the dips of 50% or so of normal values are usual and last up to a couple of seconds at the most. These disturbances (sags or swells) can be safely handled by voltage-stabilization devices, constant voltage transformers or where required, by UPS systems with battery backup. Also, most electronic power supplies are designed to ride-through short disturbances.

Surges, however, pose a more serious hazard. One of the main causes of surges is lightning. While shielding prevents direct strikes on electrical lines, induced surges cannot be altogether avoided. A lightning surge when superimposed on an AC power wave is shown in Figure 7.4. Note the sharpness of rise and the very small duration of the disturbance.

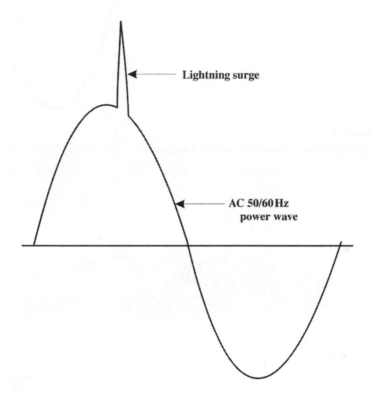

Figure 7.4
Lightning surge superimposed on AC supply waveform

The other main reason for surges is the opening of inductive loads. Magnetic energy is stored in an energized coil, which tends to continue the current flow when the circuit is broken. This gives rise to a high voltage pulse, which creates an arc across the switch contacts when they are in the process of opening. As the switch gap keeps on increasing, the arc is quenched and once again gives rise to a high voltage pulse, which can cause a re-strike. This happens a few times before the current is finally interrupted completely when the switch gap becomes too high to permit a restrike. The resulting waveform is a series of diminishing spikes superimposed over the AC sine wave (refer Figure 7.5).

Such switching surges can happen not only in large power transformers of high voltage but also in a building distribution system employing choke coils, small power supplies with transformers or relay coils used in different devices.

All these are clubbed under the name of transients or transient surge voltage. Figures 7.6 and 7.7 show some common causes of transients.

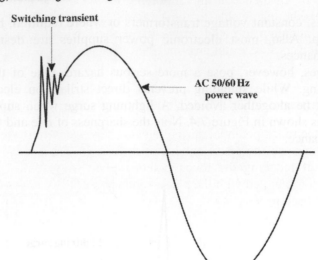

Figure 7.5
Surge superimposed on the AC supply waveform

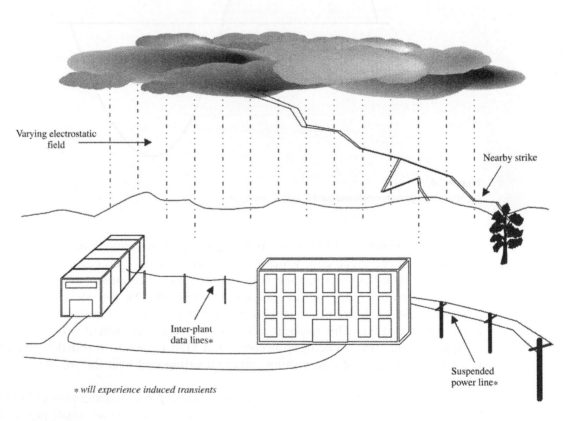

Figure 7.6
Atmospheric transients

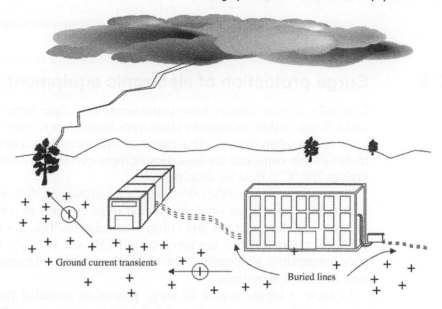

Figure 7.7
Earth current transients

7.5   Principle of surge protection

Figure 7.8 explains the basic principle involved in surge protection. A surge involves a high magnitude voltage spike superimposed on a sine wave. The transient voltage surge suppressor (TVSS) or surge protection device (SPD) is a component, which is of high impedance at normal system voltages but is conductive at higher voltages (but still below the basic insulation level of the system).

When a surge with a steep wavefront comes into the system, the portion of the wave above the breakdown voltage of the suppressor is conducted to the ground, away from the downstream equipment. The top portion of the wave is chopped of, resulting in the clipped waveform shown in Figure 7.8.

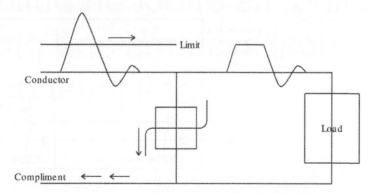

Figure 7.8
Clipping of transients

There are many devices with different voltage and power levels to suit the system which is being protected, ranging from the spark gap arrestors in power systems of high voltage to gas arrestors commonly used in communication systems. We will learn more about these devices later in this chapter.

7.6 Surge protection of electronic equipment

Generally, power circuits have components that have large thermal capacities, which make it impossible for them to attain very high temperatures quickly except during very large or long disturbances. This requires correspondingly large surge energies. Also, the materials that constitute the insulation of these components can operate at temperatures as high as 200 °C at least for short periods.

Electronic circuits, on the other hand, use components that operate at very small voltage and power levels. Even small magnitude surge currents or transient voltages are enough to cause high temperatures and voltage breakdowns. This is so because of the very small electrical clearances that are involved in PCBs and ICs (often in microns) and the very poor temperature withstanding ability of many semiconducting materials, which form the core of these components.

As such, a higher degree of surge protection is called for if these devices have to operate safely in the normal electrical system environment. Thus comes the concept of surge protection zones (SPZs). According to this concept, an entire facility can be divided into zones, each with a higher level of protection and nested within one another. As we move up the SPZ scale, the surges become smaller in magnitude, and protection better.

- *Zone 0*: This is the uncontrolled zone of the external world with surge protection adequate for high-voltage power transmission and main distribution equipment.
- *Zone 1*: Controlled environment that adequately protects the electrical equipment found in a normal building distribution system.
- *Zone 2*: This zone has protection catering to electronic equipment of the more rugged variety (power electronic equipment or control devices of discrete type).
- *Zone 3*: This zone houses the most sensitive electronic equipment, and protection of highest possible order is provided (includes computer CPUs, distributed control systems, devices with ICs, etc.).

The SPZ principle is illustrated in Figure 7.9.

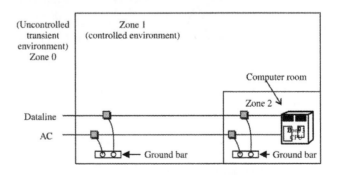

Order of magnitude of surge current	Zone 0	Zone 1	Zone 2	Zone 3
	x 1000A	x 100A	x 10A	x 1A

Figure 7.9
Zoned protection approach

We call this the zoned protection approach and we see these various zones with the appropriate reduction in the order of magnitude of the surge current, as we go down further and further into the zones, into the facility itself. Notice that in the uncontrolled environment outside of our building, we would consider the amplitude of say, 1000 A. As we move into the first level of controlled environment, called zone 1, we would get a reduction by a factor of 10 to possibly 100 A of surge capability. As we move into a more specific location, zone 2, perhaps a computer room or a room where various sensitive hardware exist, we find another reduction by a factor of 10. Finally, within the equipment itself, we may find another reduction by a factor of 10, the effect of this surge being basically one ampere at the device itself. The IEEE C62.41 indicates a similar but slightly differing approach to protection zones. The same is shown in Appendix C to this book.

The idea of the zone protection approach is to utilize the inductive capacity of the facility, namely the wiring, to help attenuate the surge current magnitude, as we go further and further away from the service entrance to the facility.

The transition between zones 0 and 1 is further elaborated in Figure 7.10. Here we have a detailed picture of the entrance into the building where the telecommunications, data communications and the power supply wires all enter from the outside to the first protected zone. Notice that the SPD is basically stripping any transient phenomena on any of these metallic wires, referencing all of this to the common service entrance earth even as it is attached to the metallic water piping system.

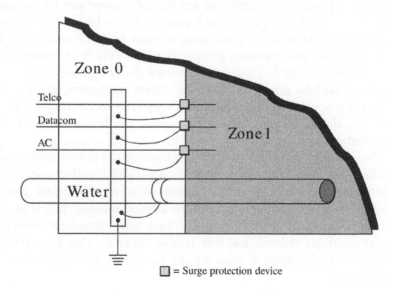

Figure 7.10
The transition from zone 0 to zone 1

Similarly, the protection for zone 2 at the transition point from zone 1 is shown in Figure 7.11.

Here as we address the discrete level between the first level of controlled zone 1 and perhaps the plug-in device taking it into the zone 2 location, we can see surge protection devices are available that handle the telecommunications, data and different types of physical plug connections for each, including both the RJ type of telephone plug as well as coaxial wiring.

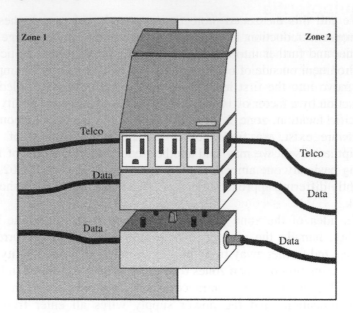

Figure 7.11
The transition from zone 1 to zone 2

This is a common design error where there are two points of entry and therefore two earthing points are established for the AC power and telecommunication circuits. The use of the TVSS devices at each point is highly beneficial in controlling the line-to-line and line-to-earth surge conditions at each point of entry, but the arrangement cannot perform this task between points of entry. This is of paramount importance since the victim equipment is connected between the two points. Hence, a common-mode surge current will be driven through the victim equipment between the two circuits despite the presence of the much-needed TVSS. The minimal result of the above is corruption of the data and maximally, there may be fire and shock hazard involved at the equipment.

No matter what kind of TVSS is used in the above arrangement nor how many and what kind of additional individual, dedicated earthing wires, etc. are used, the stated problem will remain much as discussed above. Wires all possess self-inductance and because of $-e = L \ dI/dT$ conditions cannot equalize potential across themselves under normal impulse/surge conditions. Such wires may self-resonate in quarter-waves and odd-multiples thereof, and this is also harmful. This also applies to metal pipes, steel beams, etc. Earthing to these nearby items may be needed to avoid lightning side-flash, however.

7.7   Achieving graded surge protection

From the above, it will be clear that the type of surge protection depends on the type of zone and the equipment to be protected. We will further illustrate this by a few examples, as we proceed from the uncontrolled area of zone 0. Let us begin by talking about what happens when a lightning strike hits an overhead distribution line.

Here in Figure 7.12, we see the picture of the thunderstorm cloud discharging onto the distribution line and the points of application of a lightning arrestor by the power company at points #1 and #2. We notice that the operating voltage here is 11 000 volts on the primary line and the transformer has a secondary voltage of 380/400 V typically serving the consumer. We need to understand what is known as traveling wave phenomena. When the lightning strike hits the power line, the power line's inherent construction makes it capable to withstand as much as 95 000 V for its insulation system.

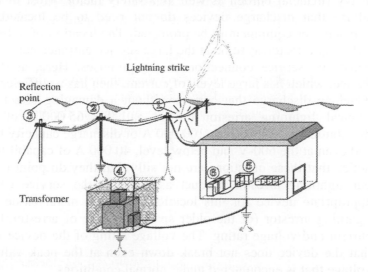

Figure 7.12
Protections in zone 0

We call this the basic impulse level (BIL). Most of the 11 000-V construction equipment would have a BIL rating of 95 kV. This says to us that the wire insulation, the cross-arms and all of the other parts, which are nearby to the current-carrying conductors, are able to withstand this high voltage.

   Traveling waves and sparks over the lightning arrestor applied on a 11 000-V line might have a spark-over characteristic of approximately 22 000 V. This high level of spark-over protection is to enable the lightning arrestor to wait until the peak of the 11 000-V operating wave shape is exceeded before discharging the energy into the earth. The peak of the 11 000-V RMS wave would be somewhere in the neighborhood of 15 000 V. As the voltage comes to the 22 000-V level and then stays there as the lightning arrestor performs its discharge, that voltage waveform travels on the power line moving very fast to all points of the line. At places where there is discontinuity to the electric line, such as points #3 or #4 in our chart, the traveling wave will go in at 22 000 V and then will double and start back down the line at 44 000 V. This type of phenomenon is known as reflection of the traveling wave and it occurs at open parts of the circuit or even the primary of transformers. When the primary of our distribution transformer serving the building achieves 44 000 V, the secondary supplying the building is going to have an over-voltage condition on it. Thus, points #5 and #6 on our chart require us to think in terms of some type of lightning-protective devices at the secondary of the transformer, the service entrance to the building and then further on into the building such as point #6 for the sensitive equipment to be fully protected in this facility.

7.8   Positioning and selection of lightning/surge arrestor

Figure 7.13 poses some serious questions. Where do we put a lightning arrestor? Do we need one at all? The final question is in answer to #1 and #2 above. What is it you wish to blow up? The computer? The UPS? And which are the most appropriate devices for protecting your equipment? It may be a humorous type of description, but the practical, real world finds many people placing lightning arrestors and discharge devices inside of occupied spaces where the equipment and the personnel represent a heavy financial burden as well as a safety factor. Most of the safety standards will advise that discharge devices do not need to be located where there are either personnel or equipment to be protected. The location of a discharge device, such as a lightning arrestor, is to be at the large service entrance earth, where the electric utility makes its service connection to the premises. Here, at this point, this discharge device, which has large levels of current, then has a sufficiently large earth plane into which to discharge that current without a damaging effect on sensitive equipment. Typical lightning arrestor ratings call for 65 000 A of discharge capacity for distribution class arrestors, 100 000 A of discharge capacity for station class arrestors, and even at the 600-V and below level, 40 000 A of capability at the minimum. So we notice that these questions are not silly, but they do point out that we need to locate our lightning arrestor product as close to the service entrance as possible. The appropriate device for this location will be a metal oxide varistor (MOV) type of lightning arrestor (or the older spark-gap type of arrestors) with the required surge current and voltage rating. The voltage rating of the device is selected in such a way that the device does not break down even at the peak value of the highest system voltage that is encountered under normal conditions.

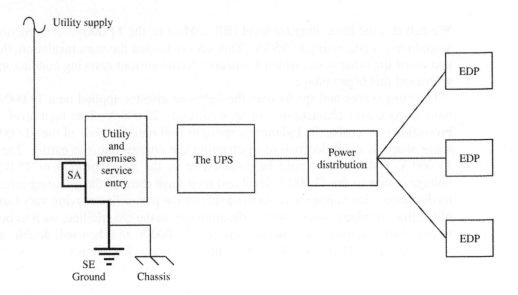

Figure 7.13
Arrestor positioning

Figure 7.14 shows the overall electric utility supply and internal wiring; we see the recommendation that is consistent with what we have been talking about before. The black boxes marked on our drawing as SPDs first appear connected at the service entrance equipment inside the building where it receives power from the service transformer. An appropriate device for this duty is the MOV type of surge arrestor. Next we see an SPD at a panel board or sub-panel assembly. Here one would preferably select a silicon avalanche device whose surge current rating may be lower but the speed for operation and low clamping voltages make it more suitable than an MOV. Finally, we may find a lower voltage style device as a discrete device either plugged in at an outlet or perhaps approaching the mounting of this device within a particular piece of sensitive equipment itself. The SPDs of silicon avalanche type are once again most appropriate in this location.

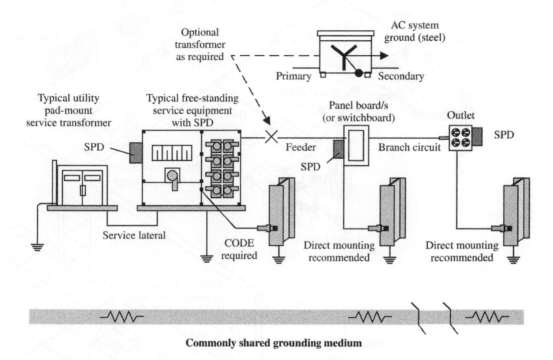

Figure 7.14
Protection locations

Figure 7.15 shows in further detail the location of SPDs in zones with sensitive equipment. Note the combined placement of lightning arrestor products and surge protection devices called transient surge protectors in this figure (reproduced from the now superceded FIPS document). Notice that the location of the arrestor product is as close to the power source as possible. In addition, also note the use of the older style arresting products, which required capacitors to affect wavefront modification. Let us explain that. Wavefront modification means that the voltage rise is so fast that if something does not mitigate that rise, the wiring may be bridged by the extremely high voltage in the surge.

Downstream from the arrestor location, over a certain amount of distance, preferably greater than 10–15 m (30–50 ft), if possible, should be the second level of protection shown in this drawing as a transient surge protector. This device indicated as combination suppressor and filter package made up of a variety of different types of components, which will now protect against the residual energy that is flowing in the circuit. The structure that we see here is one in which the various components installed in the system, starting at the service entrance, then to a sub-panel and then, finally, to discrete individual protectors, will now attenuate more and more of the surge energy until it is completely dissipated.

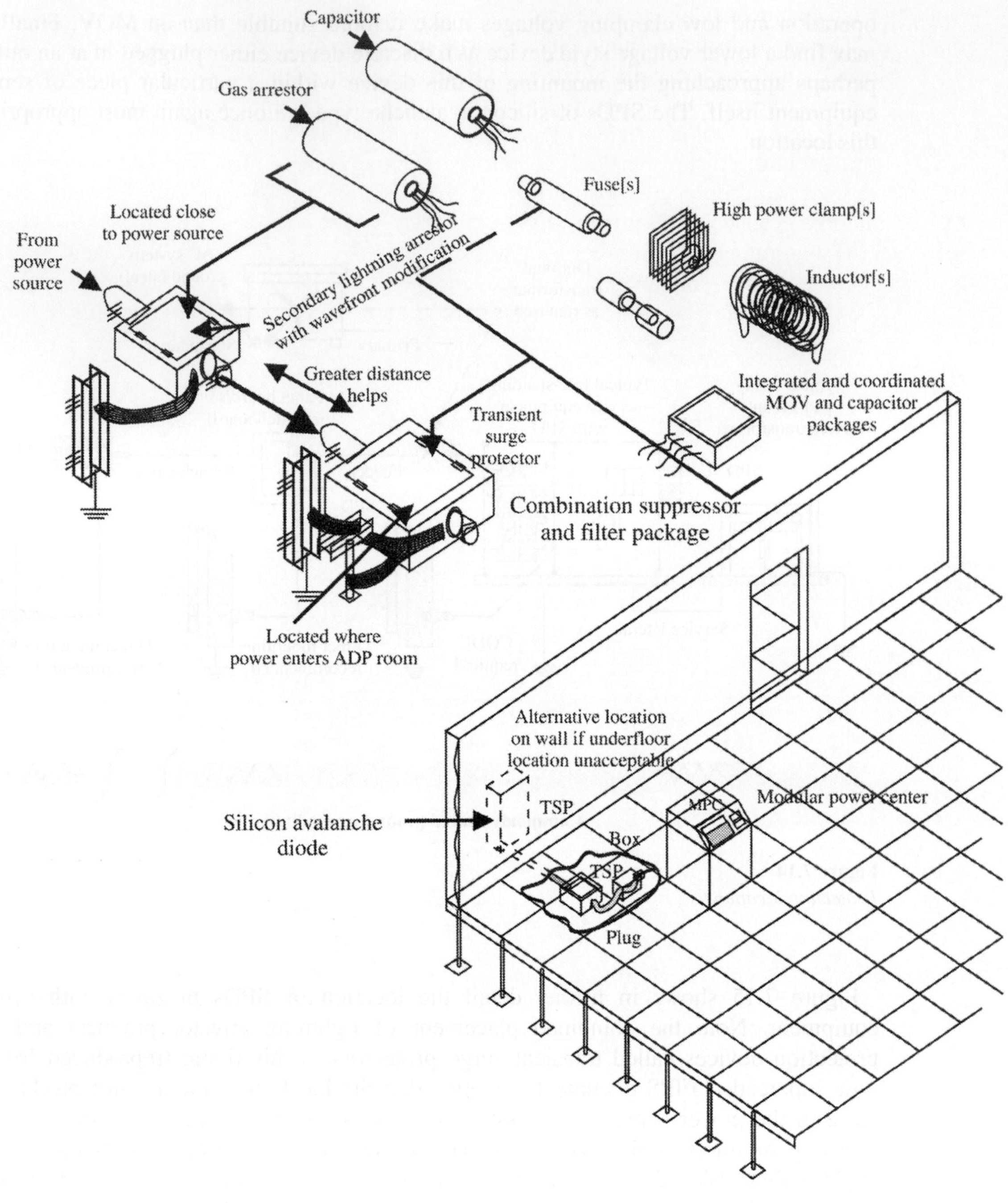

Figure 7.15
Supplementary protection

7.9 A practical view of surge protection for sensitive equipment

Any building usually has separate entry points for power and communication cables. Figure 7.16 illustrates this situation. The electrical service lateral and communication central office feeder (COF) are located at different places. Both have independent protection for surges and are separately grounded although both ground connections are inter-connected through the building's cold water piping. The sensitive electronic equipment (tagged as 'victim' equipment) has connections to the power line through a branch circuit feeder and the communication system.

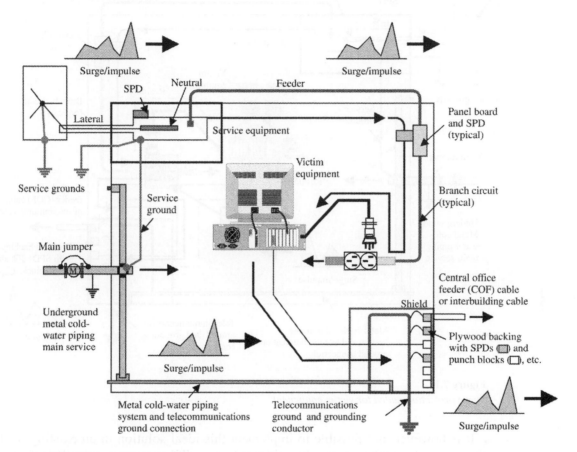

Figure 7.16
A typical site showing power and telecommunication earthing

There are two problems in this installation. The first is that the victim equipment itself has no separate surge protection and is served only by the zone 1 protection of the branch power circuit. The second problem is that surge currents flow through the building piping between the power and communication grounds, which can give rise to high-voltage differential within parts of the victim equipment.

The situation will improve somewhat by adding an SPD at the power outlet of the victim equipment (refer Figure 7.17). The problem of surge flow through the building still remains to be solved. There are two possible approaches to resolving this problem depending on whether you are planning a new facility or dealing with an existing one.

While planning a new installation, it is possible to integrate the entry points and earth connections of both power and communication services. The connection between power circuit ground and the cold-water piping is still maintained but does not cause any problem, as there is no differential voltage possible between power and communication grounding. Figure 7.18 shows such an installation.

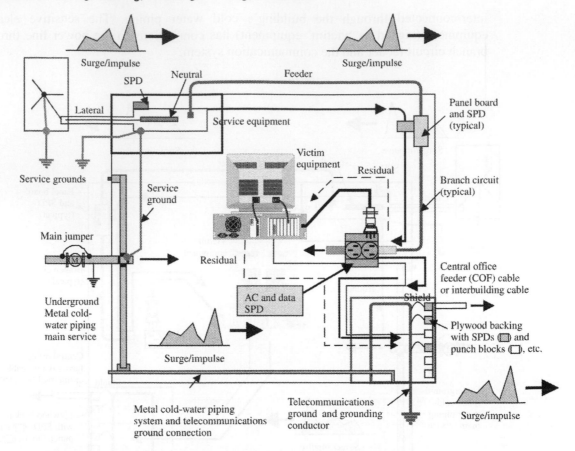

Figure 7.17
Surge protection added for victim equipment

It is however, not possible to implement this ideal solution in an existing facility. In this case, the problems can be mitigated by a different approach; that of creating a common ground plane between power and communication grounds. This is done by running a pair of metallic conduits between the power service lateral and the communication cable entry point. The communication cable is taken into the building through a new pull box with an SPD inside. The cable is then routed to the power service lateral through one of the conduits and back to the communication distribution box through the other conduit. At the power service lateral, a common ground plane is created for accommodating the communication cable loop. The victim equipment is provided with its own SPD to divert any surge currents reaching up to its power outlet. Such a system is shown in Figure 7.19.

What is best from all points of view to achieve excellent surge protection is the fully integrated facility with single ground reference plane to which all equipment enclosures and SPDs are connected and which in turn is grounded using several grounding electrodes. Such a system is shown in Figure 7.20. As far as possible, every new facility with sensitive equipment should be planned along these lines.

Codes   A number of codes, recommended practices, standards and guidelines have been developed by international and national standard making bodies on this subject and are listed in Figure 7.21. These can be used to advantage by design engineers of electrical power and data systems as well as contractors who install them.

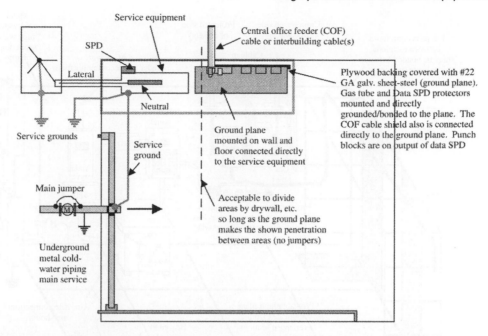

Figure 7.18

Common point of entry for power and communication cables

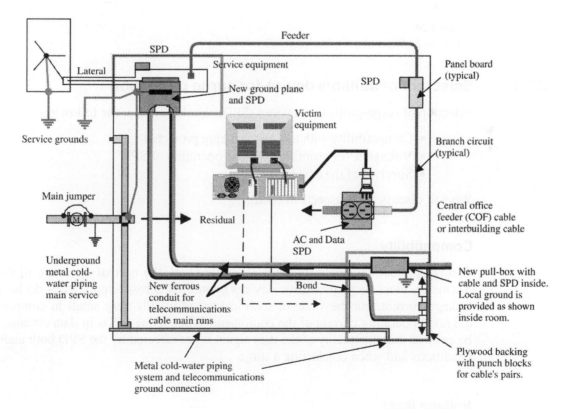

Figure 7.19

Retrofitting with a common ground plane for power and communication cabling

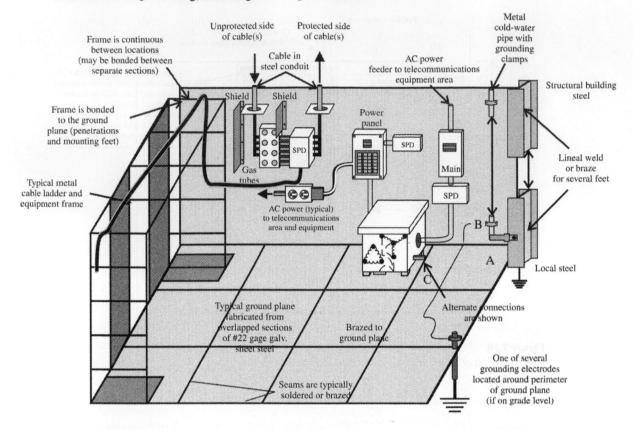

Figure 7.20
Integrated facility with a single ground reference plane

7.9.1   Selection of suitable device for surge protection

Selection of surge-protective devices (SPD) should consider the following:

- Compatibility with the system being protected
- Voltage level maintained during operation of SPD
- Survival of the device.

We will discuss these aspects in detail.

Compatibility

A surge-protective device should not interfere with the normal operation of the system. Normally, most surge protective devices connected across supply leads do have a small leakage current, but the value of such leakage is usually very small in comparison with the rated operating current of the equipment. When using SPDs in data circuits, it should be ensured that the quality of the data signal is not affected by the SPD both under normal conditions and when conducting a surge.

Voltage level

The SPD should not conduct at the normal voltage of the system (including voltage variations to which the system is normally subjected during operation). At the same time, the voltage under abnormal conditions should not be permitted to go beyond the level, which the protected system can safely withstand without any insulation breakdown.

Organization	Code, Article or Standard No.	Scope
ANSI/IEEE	C62	Guides and standards on surge protection
	C62.41 – 1980	Guide for surge voltages in low-voltage AC power circuits
	C62.1	IEEE standard for surge arrestors for AC power circuits
	C62.45 – 1987	Guide on surge testing for equipment-connected low-voltage AC power circuits
	C62.41 – 1991	Recommended practice on surge voltages connected to low-voltage AC power circuits (approved, not published)
IEEE	C74.199.6 – 1974	Monitoring of computer installations for power disturbances. International Business Machines Corp. (IBM)
UL	UL 1449	Transient voltage surge suppressors (TVSS)
NEC	Article 250	Earthing
	Article 280	Surge arrestors
	Article 645	Electronic data processing equipment
	Article 800	Communications circuits
NFPA	NFPA-75 – 1989	Protection of electronic data processing equipment
	NFPA-78 – 1989	Lightning protection code
	NFPA-20 – 1990	Centrifugal fire pumps
MIL-STD	MIL-STD-220A	50 Ω insertion loss test method. Earthing, bonding and shielding for electronic equipment and facilities
	MIL-STD-419A/B	
FIPS	FIPS PUB 94	Guideline for electrical power in ADP Installations (Chapter 7)

Figure 7.21
Codes, standards and guidelines

Survival

A surge can contain a lot of energy, which the SPD should successfully divert away from the equipment being protected. The quantum of energy varies with the location. As seen earlier in this chapter, the location can fall under different zones classified according to the severity of probable surges, with zone 0 being the worst, diminishing progressively as we move up to zone 3. The power levels of surge in zone 1 are thus the highest and zone 3 the least. An SPD selected for each one of these locations must safely clamp the voltage of the protected circuits to specified values and absorb the energy contained in the surge without permanent damage to itself. It should be remembered that lightning surges can contain multiple wave components and the SPD should withstand all of them safely. This will call for adequate energy absorption rating.

One of the important aspects of survival is that the SPD should stop conduction as soon as the surge incident is over. Some of the types of SPDs continue to conduct at relatively small voltages once the conduction mode is initiated and can therefore destroy themselves in the process unless used with other devices that can stop the conduction of current.

Importance of shielding against direct lightning strokes on power equipment. It is not always possible to design an SPD for the worst-case conditions such as a direct lightning stroke. A device that can withstand such a disturbance will be too large and prohibitively expensive to design and manufacture. The same principle will be applicable as we move up the zone of protection. The correct approach is therefore to mitigate the worst effects of surges by other means. For example, all exposed power line equipment should necessarily be shielded (refer Section 4.8). The SPD should therefore be required to protect against the residual surges only, which will make the design feasible and cost-effective.

7.9.2 Types of surge suppressors or protective devices

The IEEE 1100 contains the following description of SPDs. Various types of surge suppressors are available to limit circuit voltage. Devices vary by clamping, voltage and energy-handling ability. Typical devices are 'crow bar' types such as air gaps and gas discharge tubes, and non-linear resistive types such as thyrite valves, avalanche diodes and MOVs. Also available are active suppressors that are able to clamp or limit surges regardless of where on the power sine wave the surges occur. These devices do not significantly affect energy consumption.

Thus, there are two basic types of devices:

- Devices that limit the voltage
- Devices that switch the voltage to lower values when they break down.

The former type acts by clamping the voltage to a safe value while conducting the rest of the surge voltage and energy to the ground through the grounding lead (refer also Chapter 4 on lightning protection). It is necessary to keep the grounding lead itself as short as possible so that the voltage drop across the lead does not add to the voltage across the SPD. Metal oxide varistors (MOV as they are usually referred to) and zener (avalanche) diodes used in electronic circuits are examples of such devices. These work using the principle of non-linear resistors whose resistance falls sharply when the voltage exceeds a threshold value. But the resistance value is such that the voltage remains more or less constant even when large surge currents are conducted through the arrestor.

The second type viz. voltage-switching type (or 'crow bar' type as IEEE:1100 calls it) are devices, which suddenly switch to a low-voltage state when the voltage exceeds a certain threshold value, thus lowering the voltage 'seen' by the protected circuit. Once this happens, the normal system voltage is enough to keep the device to remain in this state. The device can be turned off only when the voltage is switched off. Spark gaps and gas arrestors are examples of such devices.

The characteristic of these devices is shown in Figures 7.22 and 7.23.

More on MOV type SPD

An MOV is essentially a non-linear resistor whose resistance to the flow of electricity varies as a function of the voltage applied to it. Zinc oxide is the basic material used in MOV type of arrestors. The arrestor element consisting of disks of zinc oxide material is kept pressed together mechanically. The diameter, thickness and number of disks determine the ratings of the device. The characteristic of an MOV resembles one shown in Figure 7.22.

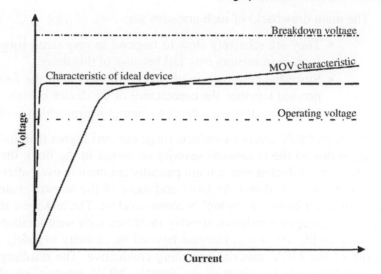

Figure 7.22
Characteristic of a voltage-limiting device (typical)

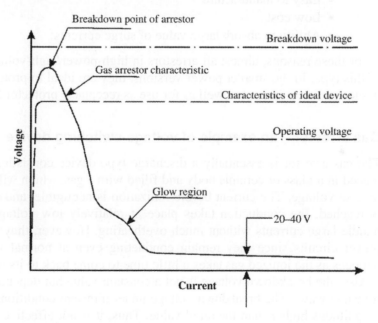

Figure 7.23
Characteristic of a voltage-switching device (typical)

Two aspects need mention. One is that there is a certain amount of leakage current even at normal operating voltage. The arrestors are designed to handle the heat loss and dissipate them safely to the environment without any danger to the arrestor. The second is that the voltage across the arrestor rises with increasing value of surge currents and thus presents a danger that the protection may become ineffective beyond a point.

The main drawbacks of such arrestors are:

- They are relatively slow to respond to very steep surges, and sensitive devices such as transistors may fail because of this delay.
- Since the construction of MOV arrestors is in the form of disks of zinc oxide pressed together, the capacitance of the device is high.
- MOV arrestors are subject to 'aging' a term which is elaborated below.

When an MOV arrestor conducts surge currents higher than its design rating (which can happen due to the occasional severity of surges in the line), the path through which the surges get conducted may remain partially conductive even after the surge passes off and the arrestor cools down. As more and more of the arrestor material is thus affected, the phenomenon known as 'aging' becomes evident. The leakage current in the arrestor under normal operating conditions steadily increases with such incidents, which causes internal heating in the arrestor to increase beyond its capacity to safely dissipate. This results in more of the MOV material becoming conductive. The resulting thermal runaway effect eventually leads to failure of the arrestor. MOV arrestors are thus protected with fuses combined with other indication to warn about the failure of the arrestor.

MOV arrestors have advantages over other types, which make them ideal in many situations. These advantages are:

- Relatively simple in construction
- Easy to manufacture
- Low cost
- Ability to absorb large value of surge currents.

For these reasons, almost all arrestors in high-power/high-voltage electrical circuits are of this type. In the smaller power versions, they are ideal to protect power supply circuits of electronic equipment as well as for use as receptacle protector SPDs.

Gas arrestor – an example of voltage-switching device

The gas arrestor is essentially a discharge type device consisting of a pair of electrodes placed in a glass or ceramic body and filled with a gas, which will ionize and conduct at a precise voltage. The current before ionization is negligible, and once breakdown voltage is reached, the conduction takes place at relatively low voltage. The gas arrestor can handle large currents without much overheating. However, they are unsuitable for use in power circuits since they remain conducting even at normal voltage once breakdown happens as the ionized gas takes a little time to come back to its normal state.

Also, the breakdown voltage is not a constant value but depends on the steepness of the transient wave. The breakdown voltage under transient conditions may be a few orders of magnitudes higher than the rated value. Thus, it is not effective for protecting electronic circuits unless used in combination with other devices.

Coordination of surge suppressors

Effective surge protection calls for coordinated action of different devices, from the large capacity current diverting devices at the service inlet, followed by a series of devices of decreasing voltage clamping and surge energy absorption ratings. The purpose of devices at the service inlet is to reduce the energy level of serious surges to values that can be handled by the downstream devices. Improper coordination can cause excessive surge energy to reach the downstream suppressors causing their failure (as well as damage to connected equipment). The principle of zones of protection explained earlier in this chapter is a practical way of obtaining this type of coordination.

7.10   Summary

In this chapter, we saw the importance of integrating the grounding systems within a building. We reviewed the causes of surges and how they can be prevented from damaging sensitive equipment. We learnt about the concept of surge protection zones and how different zones are protected. Examples from practical situations where incorrect earthing and bonding can cause problems with tips on planning an integrated facility were seen. Types of surge protection devices and their applications were also described.

8

Electrical noise and mitigation

8.1 Introduction

In this chapter, we will learn about noise in electrical circuits, the reasons for their generation, types of noise and mitigation. We will cover shielding as a means of noise control and the role played by grounding and how properly designed grounding can reduce noise. We will learn about zero signal reference grids for noise-prone installations. We will briefly deal with the subject of harmonics and how they affect power and electronic equipment and about ways of controlling them.

8.2 Definition of electrical noise and measures for noise reduction

Noise, or interference, can be defined as undesirable electrical signals, which distort or interfere with an original (or desired) signal. Noise could be transient (temporary) or constant. Unpredictable transient noise is caused, for example, by lightning. Constant noise can be due to the predictable 50 or 60 Hz AC 'hum' from power circuits or harmonic multiples of power frequency close to the data communications cable. This unpredictability makes the design of a data communications system quite challenging.

Noise can be generated from within the system itself (internal noise) or from an outside source (external noise). Examples of these types of noise are:

Internal noise

- Thermal noise (due to electron movement within the electrical circuits)
- Imperfections (in the electrical design).

External noise

- Natural origins (electrostatic interference and electrical storms)
- Electromagnetic interference (EMI) – from currents in cables
- Radio frequency interference (RFI) – from radio systems radiating signals
- Cross talk (from other cables separated by a small distance).

From a general point of view, there must be three contributing factors before an electrical noise problem can exist. These are:

1. A source of electrical noise
2. A mechanism coupling the source to the affected circuit
3. A circuit conveying the sensitive communication signals.

Typical sources of noise are devices, which produce quick changes (spikes) in voltage or current or harmonics, such as:

- Large electrical motors being switched on
- Fluorescent lighting tubes
- Solid-state converters or drive systems
- Lightning strikes
- High-voltage surges due to electrical faults
- Welding equipment.

Figure 8.1 shows a typical noise waveform and how it looks when superimposed on the power source voltage waveform.

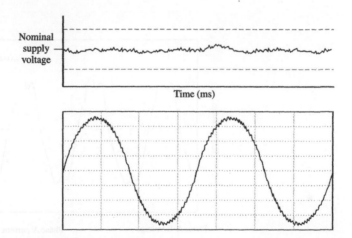

Figure 8.1
Noise signal (top) and noise over AC power (bottom)

Electrical systems are prone to such noise due to various reasons. As discussed in the previous chapter, lightning and switching surges are two of these. These surges produce high but very short duration of distortions of the voltage wave. Another common example is 'notching', which appears in circuits using silicon-controlled rectifiers (power thyristors). The switching of these devices causes sharp inverted spikes during commutation (transfer of conduction from one phase arm to the next). Figure 8.2 shows the typical waveform with this type of disturbance.

Harmonics in supply system is yet another form of disturbance. This subject will be reviewed in detail later in the chapter. A typical waveform with harmonic components is shown in Figure 8.3. Switching of large loads in power circuits to which automatic data processing (ADP) loads are connected can also cause disturbances. Similarly, faults in power systems can cause voltage disturbances. All these distortions and disturbances can find their way to sensitive electronic equipment through the power supply mains connection and cause problems. Apart from these directly communicated disturbances, sparks and arcing generated in power-switching devices and high-frequency harmonic current components can produce electromagnetic interference (EMI) in signal circuits, which will require to be properly shielded or screened to avoid interference. Figure 8.4 shows diagrammatically the reasons for noise from the equipment within a facility.

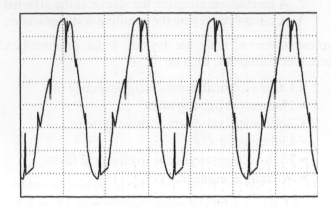

Figure 8.2
Waveform distorted by notching

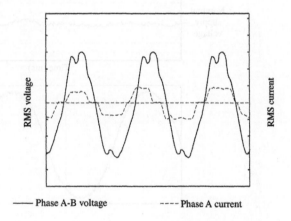

———— Phase A-B voltage   ---- Phase A current

Figure 8.3
Waveform distorted by harmonics

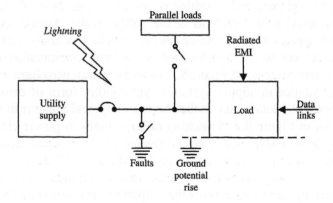

Figure 8.4
Noise emanating from electrical systems within a facility

The following general principles are applicable for reducing the effects of electrical noise:

- Physical segregation of noise sources from noise-sensitive equipment
- Electrical segregation
- Harmonic current control
- Avoiding ground loops which are a major cause of noise propagation (including measures such as zero signal reference grid, explained later in this chapter)
- Shielding/screening of noise sources and noise-susceptible equipment including use of shielded/twisted pair conductors.

8.3 How are sensitive circuits affected by noise?

Noise is only important if it is measured in relation to the communication signal, which carries the data or information. Electronic receiving circuits for digital communications have a broad voltage range, which determines whether a signal is binary bit '1' or '0'. The noise voltage has to be high enough to take the signal voltage outside these limits for errors to occur.

The power and logic voltages of present day devices have been drastically reduced and at the same time, the speed of these devices has increased with propagation times now being measured in picoseconds. While the speed of the equipment has gone up and the voltage sensitivity has gone down, the noise conditions coming from the power supply side have not reduced at all.

The best illustration that can be given of this condition is to consider where the signal voltage has been and what is happening to it compared to the noise voltage (see Figure 8.5).

In years gone by, signal voltages may have been 30 V or more but since then have steadily been decreasing. As long as the signal voltage was high and the noise voltage was only 1 V, then we had what most instrument engineers would call a very high signal to noise ratio, 30:1. Most engineers would say you have no problem distinguishing the signal as long as you have such a high signal to noise ratio. As the electronic equipment industry advanced, the signal strength went down further, below 10 and then below 5. Today we are fighting 1-, 2- and 3 V signals and still finding ourselves with 1, 2 and 3 V of electrical noise. When this takes place for brief periods of time, the noise signal may be larger than the actual signal. The sensors within the sensitive equipment turn and try to run on the noise signal itself as the predominant voltage.

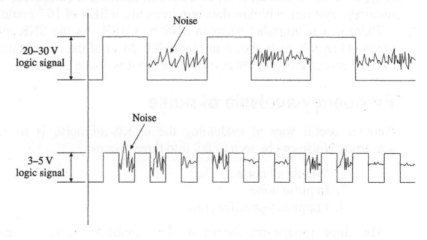

Figure 8.5
Relative magnitudes of signal and noise (then and now)

When this takes place, a parity check or a security check signal is sent out from the sensitive equipment asking if this particular voltage is one of the voltages the sensor should recognize. Usually, this check fails when it is a noise voltage rather than the proper signal that it should be looking at and the equipment shuts down because it has no signal. In other words, the equipment self-protects when there is no signal to keep it operating. When the signal to noise ratio has fallen from a positive direction to a negative direction, the equipment interprets that as the need to turn off so this it will not be running on sporadic signals.

In the top portion of Figure 8.5, a 20–30-V logic signal is well in excess of the noise that is occurring between the on and off digital signal flow. In the bottom picture, however, the noise has raised its head above the area of the logic signal which has now dropped significantly into the 3–5 V range and perhaps even lower. You will also notice that the difference between the upper and lower pictures in the graph shows the speed with which the signal was transmitted. In the upper graph, the ons and offs are relatively slow, evidenced by the large spaces between the traces. In the lower graph, the trace is now much faster. There are many more ons and offs jammed into the same space and as such, the erratic noise behavior may now interfere with the actual transmission.

The ratio of the signal voltage to the noise voltage determines the strength of the signal in relation to the noise. This 'signal to noise ratio' (SNR) is important in assessing how well the communication system will operate. In data communications, the signal voltage is relatively stable and is determined by the voltage at the source (transmitter) and the volt drop along the line due to the cable resistance (size and length). The SNR is therefore a measure of the interference on the communication link.

The SNR is usually expressed in decibels (dB), which is the logarithmic ratio of the signal voltage (S) to noise voltage (N).

$$\text{SNR} = 10\log\left(\tfrac{S}{N}\right) \text{dB}$$

An SNR of 20 dB is considered low (bad), while an SNR of 60 dB is considered high (good). The higher the SNR, the easier it is to provide acceptable performance with simpler circuitry and cheaper cabling.

In data communications, a more relevant performance measurement of the link is the bit error rate (BER). This is a measure of the number of successful bits received compared to bits that are in error. A BER of 10^{-6} means that one bit in a million will be in error and is considered poor performance on a bulk data communications system with high data rates. A BER of 10^{-12} (one error bit in a million million) is considered to be very good. Over industrial systems, with low data requirements, a BER of 10^{-4} could be quite acceptable.

There is a relationship between SNR and BER. As the SNR increases, the error rate drops off rapidly as is shown in Figure 8.6. Most of the communications systems start to provide reasonably good BERs when the SNR is above 20 dB.

8.4   Frequency analysis of noise

Another useful way of evaluating the effects of noise is to examine its frequency spectrum. Noise can be seen to fall into three groups:

1. Wideband noise
2. Impulse noise
4. Frequency-specific noise.

The three groups are shown in the simplified frequency domain as well as the conventional time domain. In this way, we can appreciate the signal's changing properties as well as viewing the amplitude in the customary time domain.

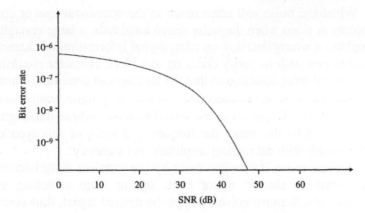

Figure 8.6
Relationship between the bit error rate and the signal to noise ratio

Wideband noise contains numerous frequency components and amplitude values. These are depicted in the time domain graph shown in Figure 8.7 and in the frequency domain graph shown in Figure 8.8.

In the frequency domain, the energy components of wideband noise extend over a wide range of frequencies (frequency spectrum).

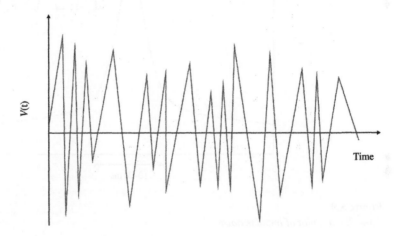

Figure 8.7
Time domain plot of wideband noise

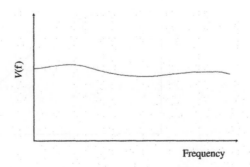

Figure 8.8
Frequency domain plot of wideband noise

Wideband noise will often result in the occasional loss or corruption of a data bit. This occurs at times when the noise signal amplitude is large enough to confuse the system into making a wrong decision on what digital information or character was received. Encoding techniques such as parity checking and block character checking (BCC) are important for wideband error detection so that the receiver can determine when an error has occurred.

Impulse noise is best described as a burst of noise, which may last for a duration of say up to 20 ms. It appears in the time domain as indicated in Figure 8.9.

Figure 8.10 illustrates the frequency domain of this type of noise. It affects a wide bandwidth with decreasing amplitude vs frequency.

Impulse noise is brought about by the transient disturbances in electrical activity such as when an electric motor starts up, or from switching elements within telephone exchanges. Impulse noise swamps the desired signal, thus corrupting a string of data bits. As a result of this effect, synchronization may be lost or the character framing may be disrupted. Noise of this nature usually results in garbled data making messages difficult to decipher. Cyclic redundancy checking (CRC) error detection techniques may be required to detect such corruption.

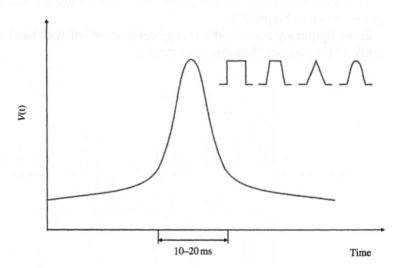

Figure 8.9
Time domain plot of impulse noise

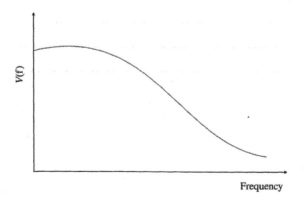

Figure 8.10
Frequency domain plot of impulse noise

Although more damaging than wideband noise, impulse noise is generally less frequent. The time and frequency domain plots for impulse noise will vary depending on the actual shape of the pulse. Pulse shapes may be square, trapezoid, triangular or sine for example.

In general, the narrower and steeper a pulse, the more energy is placed in the higher-frequency regions.

Frequency-specific noise is characterized by a constant frequency, but its amplitude may vary depending on how far the communication system is from the noise source, the amplitude of the noise signal and the shielding techniques used.

This noise group is typical of AC power systems (Figures 8.11 and 8.12) and can be reduced by separating the data communication system from the power source. As this form of noise has a predictable frequency spectrum, noise resistance is easier to implement within the system design.

Filters are typically used to reduce this to an acceptable level.

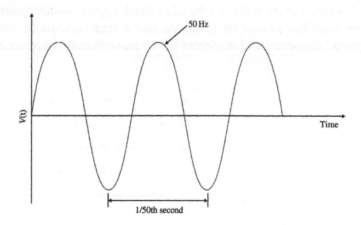

Figure 8.11
Time domain plot of constant frequency noise

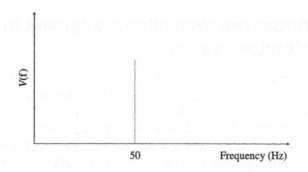

Figure 8.12
Frequency domain plot of constant frequency noise

8.5   Categories of noise

An electrical noise falls under one of the following categories: transverse mode or common mode. Transverse-mode noise is a disturbance, which appears between two active conductors (phase or neutral) in an electrical system. Such a noise is therefore measurable between two line conductors or between a line conductor and neutral. This is usually having its genesis from within the power system (Figure 8.13).

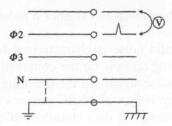

Figure 8.13
Transverse-mode noise

Common-mode noise, on the other hand, appears simultaneously in each active conductor and therefore cannot be measured like a transverse-mode noise. It usually involves the ground conductor and originates from some external disturbance (Figure 8.14).

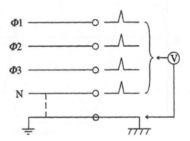

Figure 8.14
Common-mode noise

8.6 Disturbances from other equipment in the same distribution system

An important factor to be taken note of in dealing with electrical system generated noise is electrical segregation of noise-producing equipment and noise-sensitive equipment. Figure 8.15 illustrates this principle. In case A, the 'noisy' AC units and noise-sensitive ADP loads share a common power supply system. Frequent starts of AC compressors could cause voltage fluctuations, which will be communicated to ADP power units and can translate as noise in ADP units' electronic circuits. In case B, a separation of circuits has been achieved by employing different sub-circuits for AC loads and ADP loads but this may not have much impact as far as noise is concerned since the sources are shared.

In case C, a two-winding transformer has been introduced in the ADP circuit feeder. This will act as a cushion for the noise due to the inherent inductance of the transformer, which will not allow steep noise fronts to pass through. In case D, two separate transformers feed the AC loads and ADP loads with transfer-switching provision. The two-winding transformer has been retained. Obviously, D is the best case solution but expensive. In some situations, it may not be feasible to implement too. C will, however, provide an acceptable solution without being quite as expensive as D and can be retrofitted easily where required.

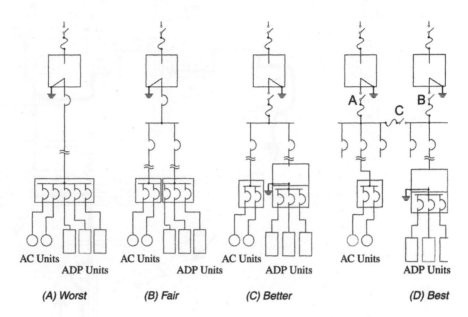

(A) Worst   **(B) Fair**   **(C) Better**   **(D) Best**

Figure 8.15
Segregation of noisy loads

8.7   Earth loop as a cause of noise

As we have seen in earlier chapters, two different equipment with a communication cable between them and each of the panels connected to a local ground point form an earth loop, which can give rise to noise currents in the communication cable. A typical building electrical system with multiple earth points is shown in Figure 8.16. Note how each panel/equipment in the distribution system is connected to ground at the nearest convenient point of the building grounding system. Note how two sensitive equipment units (shown in the upper right of the diagram as EDP devices) are connected to ground points A and B with the grounding system's inherent impedance shown between them. The EDP devices have a communication cable running between them with the ends of the cable screen connected to the EDP panel's enclosure. Any stray current in the ground system between A and B will cause a noise voltage between points A and B, which in turn can drive a current through the cable screen that can couple as a noise through the communication cable conductors.

Figure 8.17 shows how a noise can originate in the electrical power supply system. In this case, the HVAC motor winding acts as a capacitance between the electrical system and the motor's grounded enclosure. Whenever the motor starts, this capacitance sends a pulse of current through the insulation into the motor frame, which is grounded through the metallic conduit carrying the cable, leads, to the motor. The random ground connections between this conduit and other grounded metal parts act like a ground loop and create an inter-cabinet potential difference between two sensitive equipment (EDP units 1 and 2). This can cause noise pulses to flow into the serial data cable connecting the two systems, resulting in data errors.

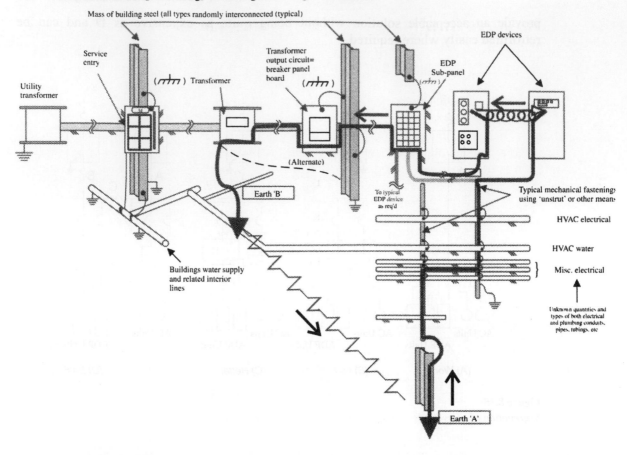

Figure 8.16
Earth connections in building electrical distribution systems causing ground loops

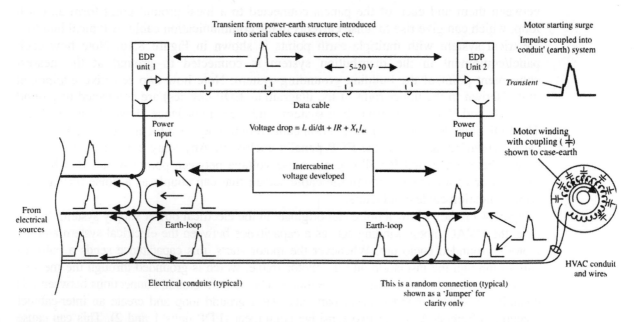

Figure 8.17
Starting of HVAC motor gives rise to noise due to ground loops

8.8 The ways in which noise can enter a signal cable and its control

Electrical noise occurs or is transmitted into a signal cable system in the following ways:

- Galvanic (direct electrical contact)
- Electrostatic coupling
- Electromagnetic induction
- Radio frequency interference (RFI).

If two signal channels within a single data cable share the same signal reference conductor (common return path), the voltage drop caused by one channel's signal in the reference conductor can appear as a noise in the other channel and will result in interference. This is called galvanic noise.

Electrostatic noise is one, which is transmitted through various capacitances present in the system such as between wires within a cable, between power and signal cables, between wires to ground (as we saw in the HVAC motor example) or between two windings of a transformer. These capacitances present low-impedance paths when noise voltages of high frequency are present. Thus noise can jump across apparently non-conducting paths and create a disturbance in signal/data circuits.

Electromagnetic interference (EMI) is caused when the flux lines of a strong magnetic field produced by a power conductor cut other nearby conductors and cause induced voltages to appear across them. When signal cables are involved in the EMI process, this causes a noise in signal circuits. This is aggravated when harmonic currents are present in the system. Higher order harmonics have much higher frequencies than the normal AC wave and result in interference particularly in communication circuits.

Radio frequency interference involves coupling of noise through radio frequency interference. We will now describe these in some detail.

8.8.1 Galvanic coupling (or common impedance coupling)

For situations where two or more electrical circuits share common conductors, there can be some coupling between different circuits with deleterious effects on the connected circuits. Essentially, this means that the signal current from one circuit proceeds back along the common conductor resulting in an error voltage along the return bus, which affects all the other signals. The error voltage is due to the capacitance, inductance and resistance in the return wire. This situation is shown in Figure 8.18.

Obviously, the quickest way to reduce the effects of impedance coupling is to minimize the impedance of the return wire. The best solution is to use a balanced circuit with separate returns for each individual signal shown in Figure 8.19.

8.8.2 Electrostatic or capacitive coupling

This form of coupling is proportional to the capacitance between the noise source and the signal wires. The magnitude of the interference depends on the rate of change of the noise voltage and the capacitance between the noise circuit and the signal circuit.

In Figure 8.20, the noise voltage is coupled into the communication signal wires through the two capacitors C_1 and C_2, and a noise voltage is produced across the resistances in the circuit. The size of the noise (or error) voltage in the signal wires is proportional to the:

- Inverse of the distance of noise voltage from each of the signal wires
- Length (and hence impedance) of the signal wires into which the noise is induced

- Amplitude (or strength) of the noise voltage
- Frequency of the noise voltage.

There are four methods for reducing the noise induced by electrostatic coupling. They are:

1. Shielding of the signal wires
2. Separating from the source of the noise
3. Reducing the amplitude of the noise voltage (and possibly the frequency)
4. Twisting of the signal wires.

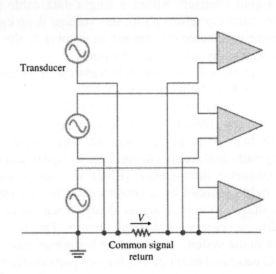

Figure 8.18
Impedance coupling

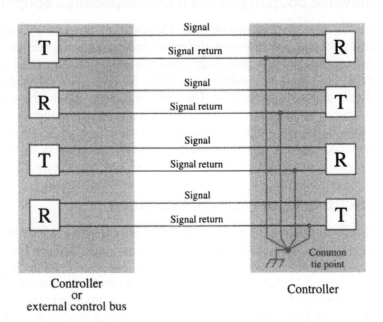

Figure 8.19
Impedance coupling eliminated with balanced circuit

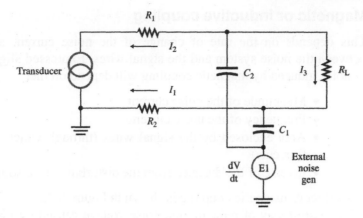

Figure 8.20
Electrostatic coupling

Figure 8.21 indicates the situation that occurs when an electrostatic shield is installed around the signal wires. The currents generated by the noise voltages prefer to flow down the lower-impedance path of the shield rather than the signal wires. If one of the signal wires and the shield are tied to the earth at one point, which ensures that the shield and the signal wires are at an identical potential, then reduced signal current flows between the signal wires and the shield.

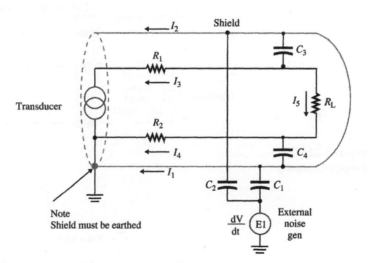

Figure 8.21
Shield to minimize electrostatic coupling

Note: The shield must be of a low-resistance material such as aluminum or copper. For a loosely braided copper shield (85% braid coverage) the screening factor is about 100 times or 20 dB that is, C_3 and C_4 are about 1/100 C_1 or C_2. For a low-resistance multi-layered screen, this screening factor can be 35 dB or 3000 times.

Twisting of the signal wires provides a slight improvement in the induced noise voltage by ensuring that C_1 and C_2 are closer together in value; thus ensuring that any noise voltages induced in the signal wires tend to cancel one another out.

Note: Provision of a shield by a cable manufacturer ensures that the capacitance between the shield and the wires is equal in value (thus eliminating any noise voltages by cancellation).

8.8.3   Magnetic or inductive coupling

This depends on the rate of change of the noise current and the mutual inductance between the noise system and the signal wires. Expressed slightly differently, the degree of noise induced by magnetic coupling will depend on the:

- Magnitude of the noise current
- Frequency of the noise current
- Area enclosed by the signal wires (through which the noise current magnetic flux cuts)
- Inverse of the distance from the disturbing noise source to the signal wires.

The effect of magnetic coupling is shown in Figure 8.22.

The easiest way of reducing the noise voltage caused by magnetic coupling is to twist the signal conductors. This results in lower noise due to the smaller area for each loop. This means less magnetic flux to cut through the loop and consequently a lower induced noise voltage. In addition, the noise voltage that is induced in each loop tends to cancel out the noise voltages from the next sequential loop. Hence an even number of loops will tend to have the noise voltages canceling each other out. It is assumed that the noise voltage is induced in equal magnitudes in each signal wire due to the twisting of the wires giving a similar separation distance from the noise voltage (see Figure 8.23).

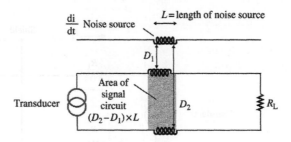

Figure 8.22
Magnetic coupling

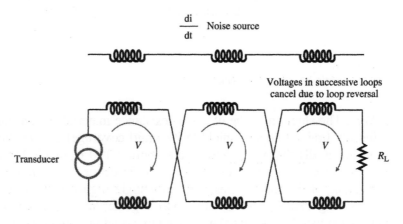

Figure 8.23
Twisting of wires to reduce magnetic coupling

The second approach is to use a magnetic shield around the signal wires (refer Figure 8.24). The magnetic flux generated from the noise currents induces small eddy currents in the magnetic shield. These eddy currents then create an opposing magnetic flux $\varnothing_1$ to the original flux $\varnothing_2$. This means a lesser flux $(\varnothing_2 - \varnothing_1)$ reaches our circuit!

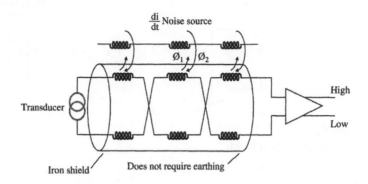

Figure 8.24
Use of magnetic shield to reduce magnetic coupling

Note: The magnetic shield does not require earthing. It works merely by being present. High-permeability steel makes best magnetic shields for special applications. However, galvanized steel conduit makes a quite effective shield.

8.8.4   Radio frequency radiation

The noise voltages induced by electrostatic and inductive coupling (discussed above) are manifestations of the near field effect, which is electromagnetic radiation close to the source of the noise. This sort of interference is often difficult to eliminate and requires close attention of grounding of the adjacent electrical circuit, and the earth connection is only effective for circuits in close proximity to the electromagnetic radiation. The effects of electromagnetic radiation can be neglected unless the field strength exceeds 1 V/m. This can be calculated by the formula:

$$\text{Field strength} = \frac{0.173\sqrt{\text{power}}}{\text{distance}}$$

where field strength is in volt/meter, power is in kilowatt and distance is in kilometer.

The two most commonly used mechanisms to minimize electromagnetic radiation are:

1. Proper shielding (iron)
2. Capacitors to shunt the noise voltages to earth.

Any incompletely shielded conductors will perform as a receiving aerial for the radio signal and hence care should be taken to ensure good shielding of any exposed wiring.

8.9   More about shielding

It is important that electrostatic shielding is only earthed at one point. More than one earth point will cause circulating currents. The shield should be insulated to prevent inadvertent contact with multiple points, which behave as earth points resulting in circulating currents. The shield should never be left floating because this would tend to allow capacitive coupling, rendering the shield useless.

Two useful techniques for isolating one circuit from the other are by the use of opto-isolation as shown in Figure 8.25, and transformer coupling as shown in Figure 8.26.

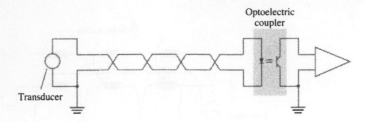

Figure 8.25
Opto-isolation of two circuits

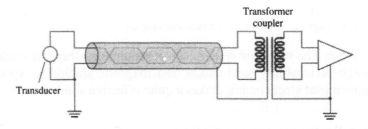

Figure 8.26
Transformer coupling

Although opto-isolation does isolate one circuit from another, it does not prevent noise or interference being transmitted from one circuit to another.

Transformer coupling can be preferable to optical isolation when there are very high speed transients in one circuit. There is some capacitive coupling between the LED and the base of the transistor which in the opto-coupler can allow these types of transients to penetrate one circuit from another. This is not the case with transformer coupling.

8.9.1　Good shielding performance ratios

The use of some form of low-resistance material covering the signal conductors is considered good shielding practice for reducing electrostatic coupling. When comparing shielding with no protection, this reduction can vary from copper braid (85% coverage), which returns a noise reduction ratio of 100:1 to aluminum Mylar tape, with drain wire, with a ratio of 6000:1.

Twisting the wires to reduce inductive coupling reduces the noise (in comparison to no twisting) by ratios varying from 14:1 (for four-inch lay) to 141:1 (for one-inch lay). In comparison, putting parallel (untwisted) wires into steel conduit only gives a noise reduction of 22:1.

On very sensitive circuits with high levels of magnetic and electrostatic coupling, the approach is to use coaxial cables. Double-shielded cable can give good results for very sensitive circuits.

Note: With double shielding, the outer shield could be earthed at multiple points to minimize radio frequency circulating loops. This distance should be set at intervals of less than 1/8th the wavelength of the radio frequency noise.

8.9.2 Cable ducting or raceways as magnetic shield

These are useful in providing a level of attenuation of electric and magnetic fields. These figures are valid for a frequency of 60 Hz for magnetic fields and 100 kHz for electric fields. Typical screening factors are:

- For 5 cm (2 in.) aluminum conduit with 0.154 in. thickness
  - Magnetic fields 1.5:1
  - Electric fields 8000:1

- Galvanized steel conduit (5 cm (2 in.), wall thickness 0.154 in.)
  - Magnetic fields 40:1
  - Electric fields 2000:1

8.9.3 Cable spacing as a means of noise mitigation

In situations where there are a large number of cables varying in voltage and current levels, the IEEE 518 – 1982 standard has developed a useful set of tables indicating separation distances for the various classes of cables. There are four classification levels of susceptibility for cables. Susceptibility, in this context, is understood to be an indication of how well the signal circuit can differentiate between the undesirable noise and required signal. It follows that a data communication physical standard such as RS-232E would have a high susceptibility, and a 1000-V, 200-A AC cable has a low susceptibility.

The four susceptibility levels defined by the IEEE 518 – 1982 standard are briefly:

1. *Level 1 – high:* This is defined as analog signals less than 50 V and digital signals less than 15 V. This would include digital logic buses and telephone circuits. Data communication cables fall into this category.
2. *Level 2 – medium:* This category includes analog signals greater than 50 V and switching circuits.
3. *Level 3 – low:* This includes switching signals greater than 50 V and analog signals greater than 50 V. Currents less than 20 A are also included in this category.
4. *Level 4 – power:* This includes voltages in the range 0–1000 V and currents in the range 20–800 A. This applies to both AC and DC circuits.

The IEEE 518 also provides for three different situations when calculating the separation distance required between the various levels of susceptibilities. In considering the specific case where one cable is a high-susceptibility cable and the other cable has a varying susceptibility, the required separation distance would vary as follows:

- Both cables contained in a separate tray
  - Level 1 to level 2–30 mm
  - Level 1 to level 3–160 mm
  - Level 1 to level 4–670 mm

- One cable contained in a tray and the other in conduit
  - Level 1 to level 2–30 mm
  - Level 1 to level 3–110 mm
  - Level 1 to level 4–460 mm

- Both cables contained in separate conduit

    - Level 1 to level 2–30 mm
    - Level 1 to level 3–80 mm
    - Level 1 to level 4 –310 mm.

The figures are approximate as the original standard is quoted in inches.

A few words need to be said about the construction of the trays and conduits. It is expected that the trays are manufactured from metal and be firmly earthed with complete continuity throughout the length of the tray. The trays should also be fully covered preventing the possibility of any area being without shielding.

Briefly galvanic noise can easily be avoided by refraining from the use of a shared signal reference conductor, in other words, keeping the two signal channels galvanically separate so that no interference takes place.

Electromagnetic induction can be minimized in several ways. One way is to put the source of electromagnetic flux within a metallic enclosure, a magnetic screen. Such a screen restricts the flow of magnetic flux from going beyond its periphery so that it cannot interfere with external conductors. A similar screen around the receptor of EMI can mitigate noise by not allowing flux lines inside its enclosure but to take a path along the plane of its surface. Physical separation between the noise source and the receptor will also reduce magnetic coupling and therefore the interference. Twisting of signal conductors is another way to reduce EMI. The polarity of induced voltage will be reversed at each twist along the length of the signal cable and will cancel out the noise voltage. These are called twisted pair cables.

Electrostatic interference can be prevented or at least minimized by the use of shields. A shield is usually made of a highly conductive material such as copper, which is placed in the path of coupling. An example is the use of a shield, which is placed around a signal conductor. When a noise voltage tries to flow across the capacitance separating two conductors, say a power and a signal conductor (actually through the insulation of the conductors), it encounters the conducting screen, which is connected to ground. The result is that the noise is diverted to ground through the shield rather than flowing through the higher impedance path to the other conductor. If the shield is not of a high conductive material, the flow of the diverted current through the shield can cause a local rise of voltage in the shield, which can cause part of the noise current to flow through the capacitance between the shield and the second conductor.

8.9.4   Optical cables

The best method is of course to use signal cables of optical type, which are immune to all forms of electrical noise. Their use is very common in communication cables and for network conductors for supervisory control and data acquisition (SCADA) systems found in major electrical installations where noise is inherent in the environment. Most industrial controls such as distributed control systems in power and various process industries prefer use of fiber optic conductors as their data highway.

All these methods are routinely applied in practice as noise reduction measures. We will discuss about one of the important components used in power distribution systems for sensitive equipment viz., the shielded isolation transformer.

8.10   Shielded isolation transformer

A shielded transformer is a two-winding transformer, usually delta–star connected and serves the following purposes:

- Voltage transformation from the distribution voltage to the equipment's utilization voltage.
- Converting a 3-wire input power to a 4-wire output thereby deriving a separate stable neutral for the power supply wiring going to sensitive equipment.
- Keeping third and its multiple harmonics away from sensitive equipment by allowing their free circulation in the delta winding.
- Softening of high-frequency noise from the input side by the natural inductance of the transformer, particularly true for higher frequency of noise for which the reactance becomes more as the frequency increases.
- Providing an electrostatic shield between the primary and the secondary windings thus avoiding transfer of surge/impulse voltages passing through inter-winding capacitance.

Figure 8.27 shows the principle involved in a shielded transformer. The construction of the transformer is such that the magnetic core forms the innermost layer, followed by the secondary winding, the electrostatic shield made of a conducting material (usually copper) and finally the primary winding. Figure 8.28 shows this detail. It can be seen that the high-frequency surge is conducted to ground through the capacitance between the primary winding (on the left) and the shield, which is connected to ground. Besides the shield, the magnetic core, the neutral of the secondary winding and the grounding wire from the electronic equipment are all terminated to a ground bar, which in turn, is connected to the power supply ground/building ground. It is also important that the primary wiring to and secondary wiring from the isolation transformer are routed through separate trays/conduits. If this is not done, the inter-cable capacitances may come into play negating the very purpose of the transformer.

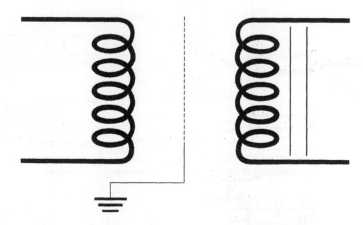

Figure 8.27
Principle of a shielded two winding transformer

Figure 8.29 shows the proper way for an isolation transformer to be wired. Note that the AC power supply wiring and the secondary wiring from the transformer are taken through separate conduits. Also, the common ground connection of the isolation transformer serves as the reference ground for the sensitive loads. The AC system ground electrode connection is taken through a separate metal conduit. If these methods are not followed and wiring/earth connections are done incorrectly, noise problems may persist in spite of the isolation transformer.

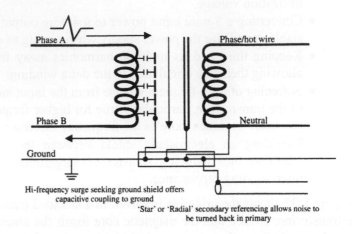

Figure 8.28
Construction of a shielded two-winding transformer

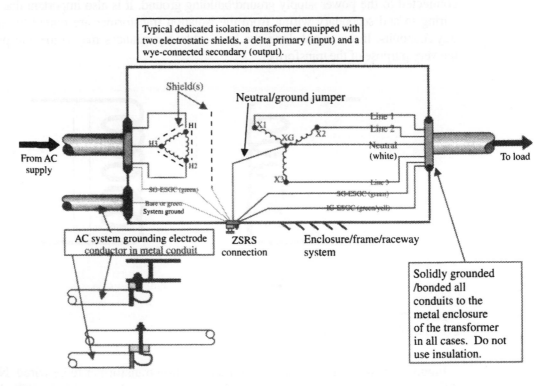

Figure 8.29
Wiring/earthing of a shielded two-winding transformer

8.11   Avoidance of earth loop

We discussed earlier in this chapter about the earth loop being a primary mechanism of noise injection into sensitive signal circuits. One of the important noise mitigation measures is therefore the avoidance of ground loops altogether. We have also seen in the previous chapters that while keeping a separate ground for the sensitive equipment may resolve noise issues, it is an unsatisfactory solution from the safety point of view.

The correct approach is therefore to keep a common electronic ground but bond it firmly with the power system ground at the source point. Figure 8.30 shows an installation with a ground loop problem.

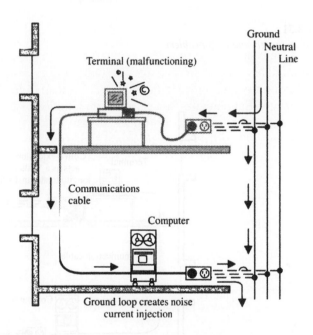

Figure 8.30
Ground loop problem

Here, the main computer system (bottom) and its user terminal are shown connected to the power circuit (including ground wiring) at two different points. A communication cable runs between the computer and its terminal. A ground loop is thus formed with the length of communication cable and the ground wire acting together in series.

Figure 8.31 shows one way in which this loop can be tackled, by bringing the two power and ground connections together to outlets at a single point.

This arrangement may not be feasible or practical to adopt. What is really possible is to introduce additional impedance in the ground loop so that the high-frequency noise prefers to take another low-impedance path and diverts itself away from the communication path. This is the principle behind the use of a longitudinal (or 'balun') transformer. Figure 8.32 demonstrates the action of this method.

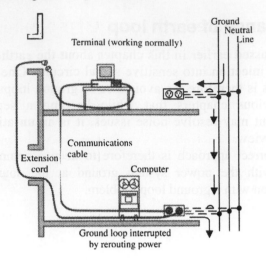

Figure 8.31
Solution to the ground loop problem

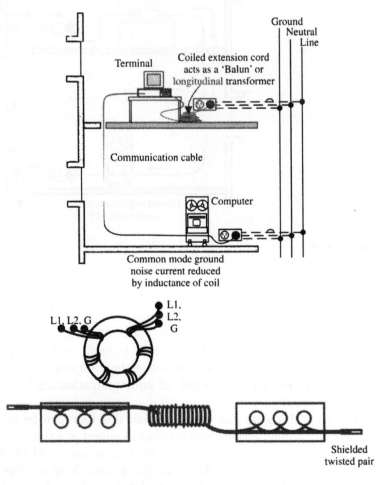

Figure 8.32
Use of a 'balun' transformer for noise mitigation

8.12   Use of insulated ground (IG) receptacle

The IG receptacles are used in situations where we wish to avoid the mixing of sensitive equipment ground and the building power system ground at all points except the power source (say, the secondary of the shielded isolation transformer) thus avoiding ground loops from forming. Figure 8.33 shows such a receptacle. The receptacle frame has a separate ground connection, which is bonded to the general ground system through the metallic conduit to ensure safe conditions. But the grounding wire from the sensitive equipment is an insulating wire, which runs through the conduit directly to the ground point of the source. Figure 8.34 illustrates such a connection.

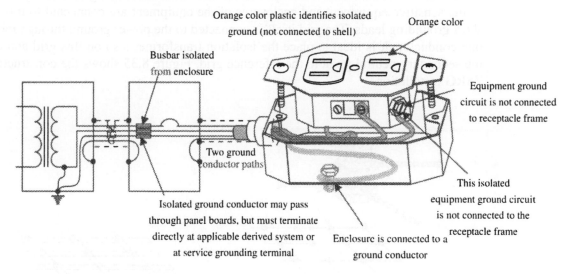

Figure 8.33
An exploded view of IG receptacle

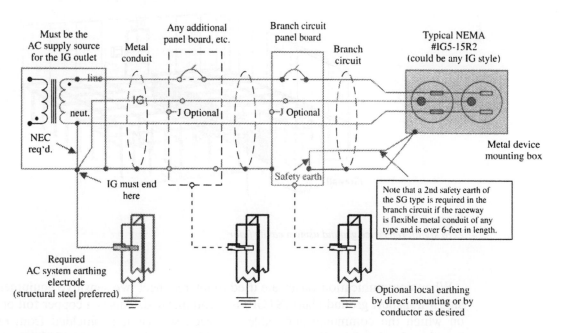

Figure 8.34
Grounding while using an IG receptacle

8.13 Zero signal reference grid and signal transport ground plane

From the foregoing, it will be clear that correct ground connection is a key factor for error-free operation of sensitive equipment and elimination of ground loops to the best possible extent is of extreme importance.

A practical way in which the above can be achieved is by using the support structures of the raised floor (which are common in computer installations and control rooms) as a ground grid called the zero signal reference grid (ZSRG). The grid is formed by the support structures of the raised floor usually arranged as 2 ft square tiles. Copper conductor of #4 AWG size is clamped to the structures forming a grid. All signal grounds of the sensitive equipment and enclosures of the equipment are connected to this grid by short grounding leads. The grid itself is connected to the power ground through more than one conductor. It is ideal to place the isolation transformer also on this grid and connect the secondary neutral point to the reference grid. Figure 8.35 shows the construction of a ZSRG installation.

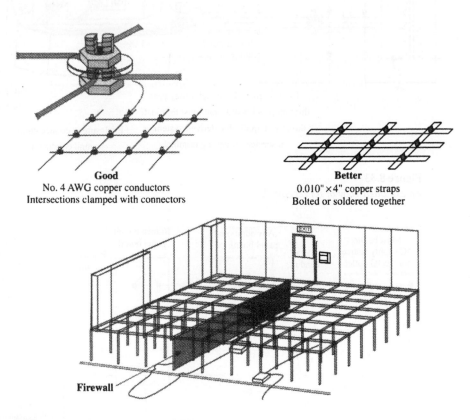

Good
No. 4 AWG copper conductors
Intersections clamped with connectors

Better
0.010" × 4" copper straps
Bolted or soldered together

Firewall

Figure 8.35
Zero signal reference ground using a cavity floor

When communication cables are used to interconnect two sensitive equipment, use of a signal transport ground plane (STGP) is recommended. This is a copper foil or a GI sheet on which the communication cable is placed so that it is shielded from electrostatic transfer of noise. Metallic cable trays on which a cable is placed and clamped close to it can also serve as an STGP. Within the same room the STGP can be bonded to the ZSRG at one or more points. When a cable runs between installations in different parts of a building, it will be necessary to have individual ZSRGs in each area and also ensure that the STGP is bonded at either end to these grids. In case balun transformers are also used on the signal cable, noise will be further reduced. Figure 8.36 shows such an installation.

Use of ZSRG has an added bonus too. It provides numerous parallel grounding paths and thus avoids resonance situation. Resonance happens when the length of a ground lead coincides with quarter wavelength of the noise frequency (or ¾, 1¼, etc.) causing the earth lead to act as an open circuit to these frequencies. With multiple ground paths, this is unlikely to happen.

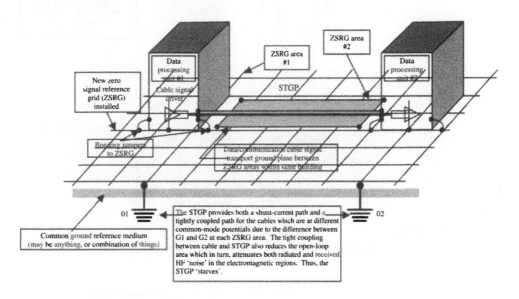

Figure 8.36
Use of signal transport ground plane

A judicious mix of ZSRG, shielded transformer and STGP used appropriately will go a long way in avoiding grounding related noise problems.

Raised floor-supporting structure as a signal reference grid:

- Bolted down stringers (struts between supporting posts) assure low electrical resistance joints.
- Isolation from building steel except via computer system earthing conductors, conductor to computer systems central earthing point and to power source earth.
- Ideal floor height for crawling access is 30 in. less than 18 in. restricts airflow. For larger computer rooms, install firewall separation barriers to confine fire and Halon extinguishing gas.

8.13.1 Self-resonance effect in grounding/bonding conductors

When dealing with grounding of power system conductors, we are concerned with the ground circuit resistance. But when dealing with circuits with high-frequency signals, it becomes necessary to consider the impedance of the grounding conductor. The grounding electrode conductors (the conductors that connect a system to the ground electrode) exhibit distributed capacitance and inductance in the length of the conductor. A particular conductor may resonate at a certain frequency (or it multiples) and may thus behave like an open-circuited conductor (refer Figure 8.37).

It is therefore advisable that grounding circuits of such systems be connected using multiple conductors with different lengths so that the combined grounding system does not resonate as a whole for any frequency. The ZSRG thus fulfills this need by providing multiple ground paths of differing lengths so that the ground path always has low impedance for any signal frequency (refer Figure 8.38).

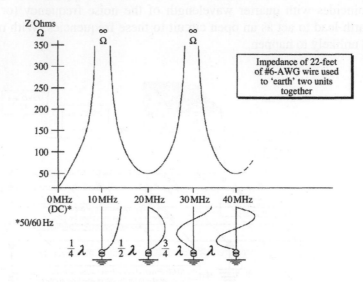

Figure 8.37
Impedance variation of a typical grounding electrode conductor

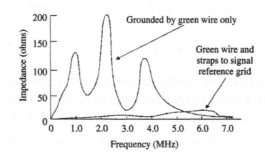

Figure 8.38
Effective grounding by ZSRG

8.14   Harmonics in electrical systems

The subject of harmonics is not directly related to this course, except that it is also a contributory factor in electrical noise. It also causes several other problems in power circuit components such as motors, transformers and capacitor banks.

A load that is purely resistive has the same wave shapes for voltage and current. Both are normally pure sinusoids. Most induction motors fed directly from AC mains also behave in a similar manner except that they draw some reactive load as well. The current waveform is still sinusoidal (see Figure 8.39).

However, the current waveform gets distorted when power electronic devices are introduced in the system to control the speed of motors. These devices chop off part of the AC waveform using thyristors or power transistors, which are used as static switches.

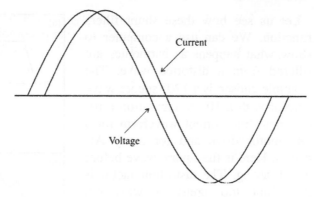

Figure 8.39
Voltage and current waveforms of an induction motor

Such altered waveforms may be mathematically analyzed using Fourier transforms as a combination of vectors of the power frequency (50/60 Hz) and others whose frequency is a multiple of the power frequency. The power frequency component is called the fundamental and higher multiples are called harmonics. It should be remembered that all electrical generators produce only voltage at fundamental frequency. But there has to be a source if a harmonic current has to flow. It is therefore construed theoretically that all harmonic-producing loads are current sources of harmonics. These sources drive harmonic currents through the rest of the system consisting of the source as well as other loads connected to it. These currents flowing through the different impedances of the system appear as harmonic voltages. It is usual for the voltage waveform of such a system to appear distorted. Also, the harmonic currents flowing through the other loads of the system give rise to several abnormalities (refer Figure 8.40 for the effects harmonics have on different system components).

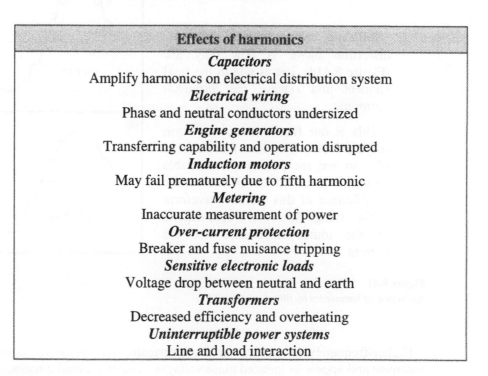

Effects of harmonics
Capacitors
Amplify harmonics on electrical distribution system
Electrical wiring
Phase and neutral conductors undersized
Engine generators
Transferring capability and operation disrupted
Induction motors
May fail prematurely due to fifth harmonic
Metering
Inaccurate measurement of power
Over-current protection
Breaker and fuse nuisance tripping
Sensitive electronic loads
Voltage drop between neutral and earth
Transformers
Decreased efficiency and overheating
Uninterruptible power systems
Line and load interaction

Figure 8.40
Effects harmonics have on different system components

Let us see how these shunt filters function. We can use a computer to show what happens as harmonics are filtered from a distorted wave. The example chosen is a 120° square wave current with a 10° commutation time; a typical line current waveform for a DC motor drive and for many AC drives. Here is the square wave before any filtering. The distortion factor is 26% not too pretty a waveform (Figure 8.41a). *Now let us take out the fifth harmonic.*

This may not look a whole lot better, but the distortion factor is down from 26 to 18%, so things are improving (Figure 8.41b). *Now let us take out the seventh as well.*

Things are actually looking better now. We can see the sine wave starting to emerge. Distortion factor is down to 11% (Figure 8.41c). *Next, we take out the eleventh.*

Still no beauty queen, but the distortion factor is now only 8% (Figure 8.41d). Let us add in the final element and *remove the thirteenth harmonic.*

This is our final current waveform (Figure 8.41e). The distortion factor is 6%, so we are putting a reasonable current into the utility. Of course, the significance of this current waveform to the voltage distortion would depend on the source impedance and the current level.

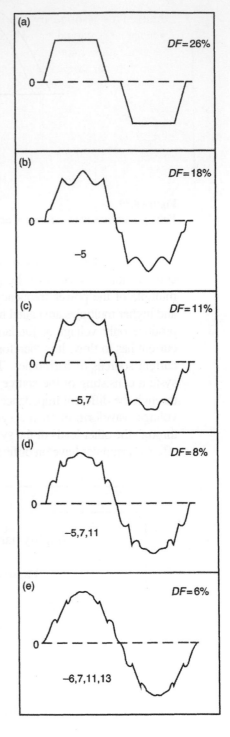

Figure 8.41
Reduction of harmonics by filters

Higher-frequency harmonics can be propagated by the power conductors acting as antennae and appear as induced noise voltages in nearby signal circuits.

It is not possible to prevent harmonic currents altogether. But they can be prevented from flowing through the entire system by providing a separate low-impedance path for them. This is done by the use of adequately rated series tuned circuits consisting of a reactor and capacitor, which have equal impedance at a specific harmonic frequency. Several such tuned banks (one for each harmonic frequency) will be needed to totally divert all harmonics away from the system. However, for practical reasons, only a few of the lower order harmonics with larger magnitudes are filtered out, which is adequate to provide substantial reduction of harmonic content.

Figure 8.41 shows how a filter might remove the high-frequency components and how the wave shape might appear as the removal takes place. A full treatment of this subject is beyond the scope of this book.

8.15   Summary

In this chapter, we have dealt with electrical noise in detail and the ways in which noise finds a path into sensitive signal circuits. We learnt the various methods by which noise can be reduced by avoiding shields, by separating the cabling, by using shielding transformers, by eliminating earth loops and by using zero signal reference grounds and signal transport ground planes. We also briefly dealt with the generation of harmonics and how they can be filtered.

9

UPS systems and their grounding practices

9.1 Introduction

Process control/SCADA systems used in industries and computers/communication equipment deployed in offices are sensitive to power interruptions and voltage/frequency excursions and require an uninterrupted power supply of good quality free from harmonics, voltage irregularities, etc. In this chapter, we will take a look at different uninterrupted power supply options available and in particular solid-state invertor-based systems. We will also touch on the issues that are important in selecting small and medium capacity UPS systems that one finds in common use these days in many premises. The issues relating to the grounding of these systems, as separately derived sources, will also be discussed from extensive references available in IEEE: 142 publication.

9.2 Power quality issues

Power supply systems in themselves cannot guarantee 100% reliability. Nor can they ensure that power supply parameters stay within the stipulated limits all the time. This is so because the point of generation and the point of usage are separated by large distances from a few miles to a few hundred miles. The power supply to consumers comes from utility grids with scores of large generators interconnected by long transmission lines, step up and step down transformers, distribution lines and switchgear at different points along the transmission and distribution chain. The transmission and distribution lines mostly are through overhead lines, which are prone to faults by lightning, failure of insulators, and so on. Even the failure of a single line can have a cascading effect by causing system instability due to sudden load shifts and can cause much more extensive disruption.

Generating equipment can also have outages due to various reasons and in turn can cause frequency excursions in the system they feed to, as it struggles to cater to the increased demand on the remaining generators. Availability of adequate spinning generation reserve mitigates the problem to a large extent.

Tripping of lines or generators can alter the flow of power in a grid system resulting in sustained outages. The process of detection and clearing of a fault is accompanied by momentary voltage drop in the affected and adjacent parts of the transmission network. Once the fault is cleared, the resulting power flow changes can cause sustained voltage drop due to overloading of the healthy lines. Disturbances are also caused by abnormalities in the distribution system closer to the consumer. As a matter of fact, close short circuit faults can cause severe voltage dips in the consumer supplies and can cause resetting of computer hardware since their switched mode power supply cannot ride through all but very brief voltage dips. Motors fed from the affected distribution systems can be disrupted, as the contactors feeding them tend to drop out during these dips. Also, sometimes the motors themselves pull out of step, as the torque generated during reduced voltage situations cannot sustain the torque required by the mechanical loads they drive.

In this chapter, we will mainly consider voltage amplitude disturbances, which can be corrected by UPS systems. Frequency disturbances and harmonic voltages require other corrective equipment, which are beyond the scope of this discussion.

Figure 9.1 shows the results of a study by EC&M magazine regarding occurrences of voltage disturbances and gives us an idea with the frequency of such occurrences in any supply network.

Disturbance	Average monthly occurrences	Percent	Notes
Oscillatory (decaying transients)	62.6	48.8	<16 ms (1 cycle) duration and >15% of system voltage
Voltage spikes	50.7	39.5	>15% of system voltage
Under voltage	14.4	11.2	<8 ms (1/2 cycle) duration and >10% of system voltage
Over voltage	None	–	<8 ms (1/2 cycle) duration and >10% of system voltage
Total outage (blackout)	0.6	0.5	Mean outage 11 min
	126.3	**100.0**	

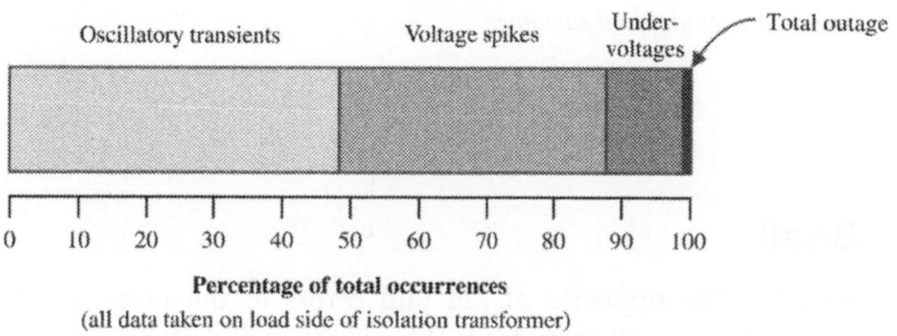

Figure 9.1
Power supply disturbances – a breakup

9.3   Definitions of abnormal voltage conditions

9.3.1   Sag

A sag is a temporary reduction in the normal AC voltage. A momentary sag is a variation, which lasts for a period of 0.5 cycle to about 2 s usually the result of a short circuit somewhere in the power system. Instances of longer duration of low voltage are called sustained sags (see Figure 9.2).

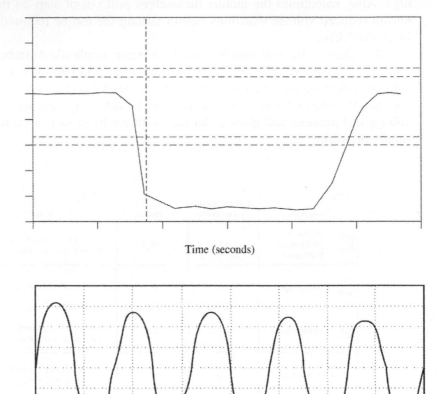

Time (seconds)

Figure 9.2
Sag – momentary and sustained

9.3.2   Swell

Swell is the opposite of sag and refers to the increase of power frequency voltage. A momentary swell lasts from 0.5 cycles to 2 s. A sustained swell lasts for longer periods (see Figure 9.3).

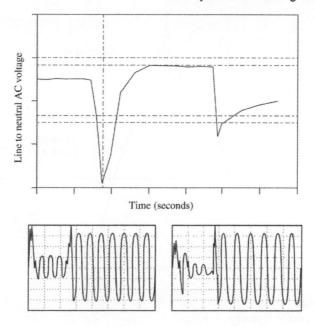

Figure 9.3
Swell – momentary and sustained

9.3.3   Surge

Surge is a sub-cycle disturbance lasting for a duration of less than half a cycle and mostly less than a millisecond. The earlier terminology was transient or spikes. The decay is usually oscillatory. Surges generally occur due to atmospheric disturbances such as lightning or due to switching of large transformers, inductors or capacitors (see Figures 9.4a and b for examples).

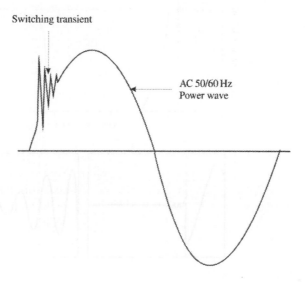

Figure 9.4a
Surge voltage with oscillatory decay

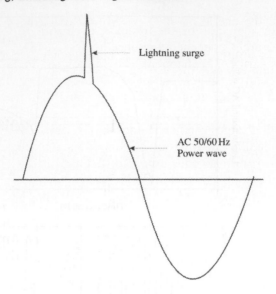

Figure 9.4b
Surge caused by lightning

9.3.4   Interruption

Interruption means the complete loss of voltage. A momentary interruption lasts from half-cycle period to less than 2 s. Longer interruptions are called sustained interruption. Momentary interruption is usually the result of a line outage with the supply being restored automatically from another source or by auto-reclosing operation. Refer Figure 9.5 for illustration. An interruption can be instantaneous or of slowly decaying type.

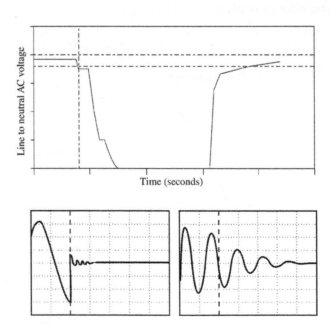

Figure 9.5
Examples of supply interruption

In Figure 9.5, the one at the top shows the RMS voltage value during a momentary interruption. The figure on the lower left depicts the waveform of a sustained interruption where the voltage drops to zero almost instantaneously. The waveform on the lower right shows an interruption where the voltage decays slowly.

9.4   Susceptibility and measures to handle voltage abnormalities

Many modern office equipment including desktop computer systems can tolerate voltage fluctuations to some extent by virtue of large capacitances (in relation to load currents) and internal regulation circuitry and can ride through a voltage sag or even a momentary interruption. The tolerance range for voltage fluctuations can be −13 to +6% for slow/sustained variations (sag/swells) and greater for short time disturbances. However, longer disturbances call for measures to counter their effects and apply corrections.

Sustained low voltages can be corrected by voltage-regulating transformers. Interruptions on the other hand can be tackled by standby sources. Since engine-operated power sources generally need time to start up and pick up load some other type of source, which can feed the load without there being a break of supply will be required. Both mechanical and electrical no-break sources are available in the market and can be used depending on applications.

We will briefly review these devices in the next paragraphs.

9.5   Regulating transformer

Slow RMS voltage variations can be effectively tackled by regulating transformers or transformers with on-load tap changers. These devices can be made to correct voltage fluctuations using voltage sensors and circuitry to effect corrections. One type of regulating transformer (a single-phase variety) is shown in Figure 9.6.

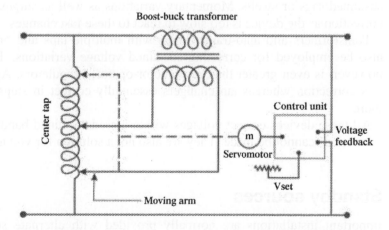

Figure 9.6
Regulating transformer

In this device, a continuously variable auto-transformer with a center tap is connected to the primary of a boost-buck correction transformer. The variable arm of the auto-transformer is operated by a servomotor, which is driven by a control circuit which compares the output voltage against a set point value. The amount and polarity of correction is decided by the voltage feedback. Such devices are also called as servo stabilizers.

Three-phase systems employ a similar device called an induction regulator. A typical scheme is shown in Figure 9.7.

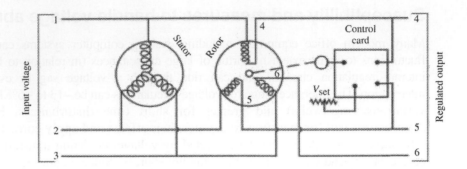

Figure 9.7
Three-phase induction regulator

This device is a transformer constructed like an induction motor. The stator of the device is connected to the power source and forms the primary of the regulating transformer. The secondary winding, which is similar to a wound rotor of an induction motor is connected in series with the supply. The secondary voltage induced in the rotor is of constant magnitude but the phase angle varies with the position of the rotor vis-à-vis the stator. The regulating circuit senses the voltage output, compares it with a set value and sends its output to a servo-motor. The servo-motor adjusts the rotor position based on the command from the regulating circuit. The output voltage is thus adjusted to the set value.

The vector diagram shown in Figure 9.8 explains how regulation is obtained.

The induction regulator is essentially an electromechanical device and can correct sustained sags or swells. Momentary variations as well as surges are let through without correction as the device is too slow to react to these fast changes.

Transformers and auto-transformers with multiple taps and on-load tap changers can also be employed for correcting sustained voltage variations. Their time of operation, however, is even greater than servomotor-operated regulators. Also, regulators give step less correction whereas tap changers essentially correct in steps of the order of 1% or more.

All these devices correct voltages within a predetermined band of limits beyond which corrections cannot be made. They are also not a solution for voltage interruptions.

9.6   Standby sources

Important installations are normally provided with alternate supply source to ensure continuity of power supply in the event of failure of the normal source. This can be in the form of a second power feeder or a standby source itself. Refer Figure 9.9, which illustrates this arrangement.

In the first case of a standby or duplicate feeder as it is sometimes called, the chance of a total failure cannot be ruled out as both feeders may be from a common source of the same power utility and therefore may fail simultaneously. The second case with an independent standby source is more reliable.

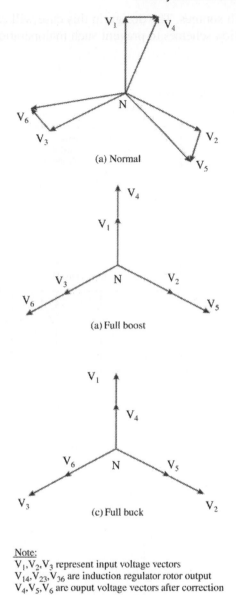

(a) Normal

(a) Full boost

(c) Full buck

Note:
V_1, V_2, V_3 represent input voltage vectors
V_{14}, V_{23}, V_{36} are induction regulator rotor output
V_4, V_5, V_6 are ouput voltage vectors after correction

Figure 9.8
Vector diagram of induction regulator

It should however be noted that the changeover from normal source to standby source involves a delay of up to 1 s even though this is done using automatic circuit breakers. In the case of the independent source, which is usually an engine-driven alternator, several seconds may elapse before the engine starts and takes over the load. Thus, they are not useful where even momentary interruptions cannot be tolerated.

It is possible to run the sources in parallel. In the case of the independent source option, this will mean running the alternator synchronized with the utility power source. Paralleled source configurations have to be designed carefully to ensure that both sources do not trip in the event of short circuits or other faults. Unless this is achieved, the use of standby sources cannot ensure uninterrupted power.

Figure 9.10 with duplicate incoming feeders illustrates how a fault F in one of the incoming lines causes short circuit currents to flow in both feeders and is likely to cause tripping of both sources. The design in this case will call for time-coordinated directional current protection schemes to prevent such maloperation.

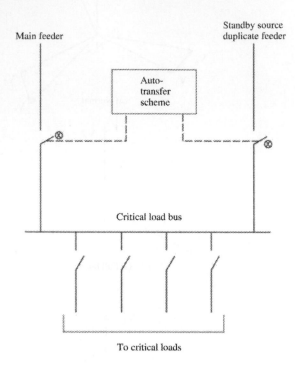

Figure 9.9
Standby power source

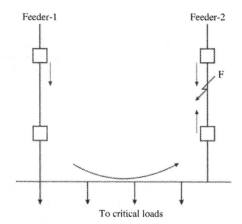

Figure 9.10
Current flow in duplicate feeders under fault conditions

Figure 9.11 illustrates a case of using a standby source, which is synchronized with the utility source and sharing the load. In case of a major failure in the utility system, the generator has to take the full load and therefore can get overloaded. Under certain conditions the standby source may send power back into the utility grid, causing the generator and engine to sink due to heavy load application all of a sudden. Such an eventuality has to be protected using relays to sense low frequency/rate-of-change of frequency or reverse power flow or both. Also, some of the in-house loads which are less critical may have to be disconnected to relieve the load on the generator.

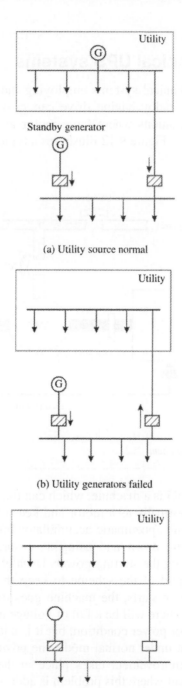

(a) Utility source normal

(b) Utility generators failed

(c) Utility feed isolated some
loads disconnected

Figure 9.11
Load flow in standby engine source under utility fault condition

Such elaborate systems are possible only in large industrial plant power systems. Even in these systems problems could occur in downstream distribution, which can disrupt power to critical LV loads. Other measures at LV level are therefore necessary to ensure uninterrupted power supply to critical equipment. Such uninterrupted systems can be electromechanical or electronic. We will discuss them in detail in the coming paragraphs.

9.7   Electromechanical UPS systems

Various electromechanical systems employing motor generators with an energy storage mechanism and a standby engine drive can provide an excellent uninterrupted power supply source in situations where large motor and electronic loads of several hundred kilowatts is to be fed. Figure 9.12 illustrates a typical system.

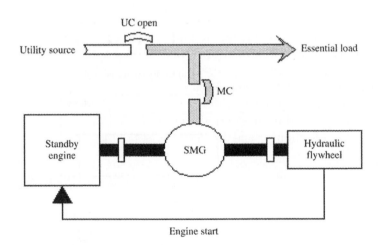

Figure 9.12
Electromechanical UPS system

In this system, SMG is a machine, which can run both as a generator and a synchronous motor. It operates normally as a motor and keeps the energy storage system, normally a flywheel or a hydraulic/pneumatic accumulator fully charged. When the power fails, the system starts slowing down and using a sensor, the engine can be started with the accumulator providing the starting power. Usually an electromagnetic clutch (not shown in the diagram) is part of the scheme to keep the engine disconnected from the SMG shaft. Once the engine starts, the machine goes to alternator mode and starts supplying power to the loads. There will be a fall of voltage and frequency due to the slowing down of SMG under loss of power condition, but it is unavoidable in this configuration. Also, it should be noted that under normal mode, the power comes from the primary source and no power quality improvement takes place in the supply system as such. Figure 9.13 shows an arrangement where this problem is addressed.

This system is similar to the earlier one except for an additional generator mounted on the same shaft for feeding to critical loads. The critical loads are thus completely isolated from the external system with the power quality solely dependent on the generator itself. A bypass switch is provided to enable power from the normal source should there be any problem with the isolated generator.

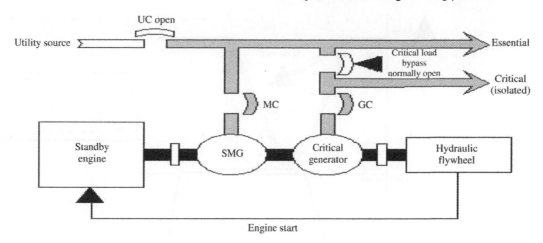

Figure 9.13
Electromechanical UPS system – a variant for critical loads

Another version of the combined type of system represented in the case that we just looked at is the dynamic diesel continuous power system as shown on these diagrams from Holec Corporation (Figure 9.14). We are dealing with a system, which has critical power, but without storage battery reliance, and utilizes a combined diesel engine generator driving through an inner rotor/outer rotor clutch system to drive an output generator to serve the load.

When we look at the results of the speed/time characteristic of this continuous power system we find that while the diesel engine remains off, the inner rotor of the induction coupling is running at full 5400 rpm. Then there comes a changeover to emergency operation where the energy stored in the rotation at 5400 rpm allows the outer rotor to drive the generator set continuously at 1800 rpm while the diesel comes up to fill in the gap. The emergency operation takes place with the diesel running and then a return to normal operation occurs after that (refer Figure 9.15, which illustrates this principle).

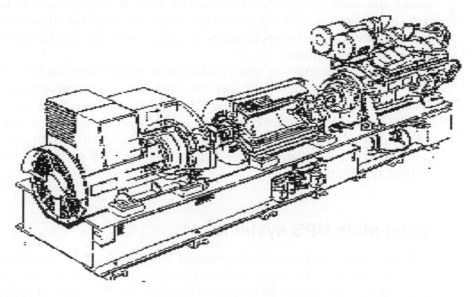

Figure 9.14
Continuous power from a system deploying electrical rotary clutch

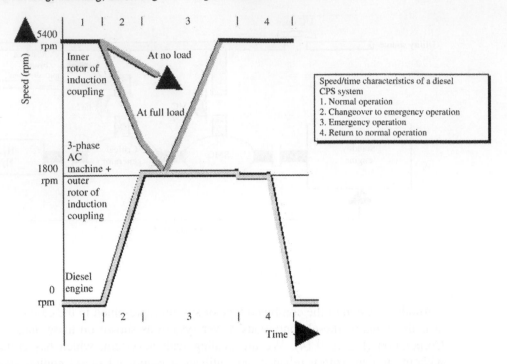

Figure 9.15
Speed time characteristic of engine-based UPS system

These engine-based UPS systems are usually preferred for very high capacity requirements amounting to several hundreds of kilowatts. The main advantage of the system is that the maintenance and replacement requirements normally associated with large battery banks are totally absent. Also, the harmonic generation by static UPS systems is not an issue when using rotary UPS systems. The maintenance requirements of engine and rotating parts are specific to the rotary type and the power losses and noise are problems to contend with. It is usual to find these systems in industrial plants producing man-made fibers to feed the fiber line drives. Some manufacturers also cite applications in critical military or aviation systems such as large radar installations.

For other applications in small and medium power range including supply to large ADP installations, static UPS systems with battery power source have become a widely deployed option. A detailed discussion on these systems follows. Also of special interest to us in this book is the grounding requirement for static UPS systems, which depends on the type of configuration employed. Since electrical noise is a problem to contend with in ADP installations, the UPS grounding issue has received critical attention.

9.8   Solid-state UPS systems

In Figure 9.16, we have the conventional block diagram for a solid-state UPS system consisting of the AC input to a rectifying device, battery bank, which forms the emergency power source, an invertor, which converts the DC to an AC sine wave output. The rectifying device supplies the DC input to the invertor and also charges the battery bank under normal conditions.

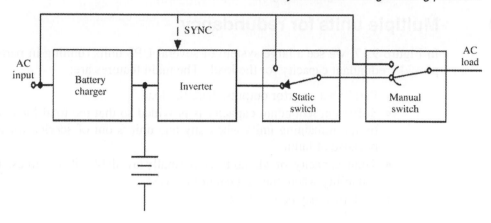

Figure 9.16
Basic solid-state UPS system

The battery storage itself is attached between the charger and the invertor. Then the invertor system converts DC back into AC, sends the AC through a static switch and a manual maintenance switch out to the AC load. The inverter, as you can see, is synchronized with the input power frequency and also has a bypass line to the static switch in the event of some type of problem with the rectifier and invertor circuit.

UPS systems can either be of on-line or off-line variety. In conventional on-line technology one would have the static switch, as shown, where the AC input power, the DC output from the battery charger and the AC output of the inverter all stay on-line continuously. The static switch is not needed to operate, unless there is a severe inrush on the load, a fuse blowing, circuit breaker operation on the load or some problem with the actual output of the invertor part of the system. This is, by far, one of the more popular versions in which no switching takes place when the AC power from the utility is no longer available. In the case of lost input power, the battery charger simply no longer functions, the invertor looks to the DC bus maintained by the batteries and takes its energy from the batteries without any further switching operation.

As an alternative, sometimes in smaller sizes and for less expense than the on-line system, there are other versions where the static switch is actually on the bypass. This is the off-line variety of UPS.

This saves a good deal of the energy conversion that goes on, but gives rise to some questions as to what type of operating characteristics one can expect when there is nothing between the outside power source and the load that you are protecting. Many of the smaller systems can be built this way, in as much as the load that they are protecting does not worry about any type of power conditioning or separation from the outside, but merely is looking for the uninterruptibility of the battery-supported system. One area to watch for in off-line systems is the fact that they have a tendency to switch rather frequently back and forth. This sometimes creates disturbances because the AC input voltage rises and falls and there is no further regulation of that voltage as the system operates.

Large UPS installations seldom use the off-line model. As a matter of fact, redundant invertor modules sharing the load with a separate bypass supply operating through a static switch is the configuration that most critical installations use.

9.9 Multiple units for redundancy

In Figure 9.17, we see a large system of several UPS units running in parallel in order to provide redundant capacity for the load. The main features are:

- Paralleled invertor outputs share the load.
- Sufficient redundant capacity is provided so that the total load will be supplied by the remaining units when any one unit is out of service for maintenance or because of failure.
- Total capacity of all units is normally available at all times, but at reduced reliability when one unit cannot be used.
- Batteries may be paralleled.

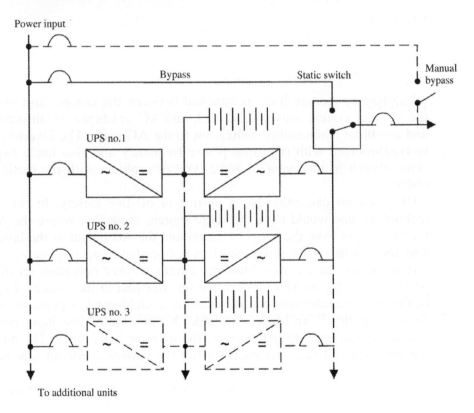

Figure 9.17
Solid-state UPS system in multiple redundant configuration

Note that in this arrangement, we do not need the bypass line as a regular function, but will always have one additional unit more than necessary to run the load. In this example, there might be three 100-kVA units where the load requires 200 kVA and all three units are running at approximately 65–70% capacity. Should one unit fail, a small load step is effected within the range of the specification of the units and the load continues to run without any switching or without transfer to bypass. It continues to maintain the protected element of the circuit.

Though in the initial days of UPS systems, the invertor modules were designed using thyristor elements, the advent of the insulated gate bipolar transistor (IGBT) has all but replaced invertor grade thyristors as well as gate turn off (GTO) devices. Being functionally similar to transistors, they offer design simplicity, faster switching, lower losses and produce less audible noise. The IGBTs in high-frequency pulse width modulated type configuration is the preferred choice of today's invertor designers.

9.10 Considerations in selection of UPS systems for ADP facilities

Designers of facilities for large ADP installations should keep in mind a few basic facts regarding the loads, which they feed and the demands they make on the power supply equipment including the UPS.

The power supply infrastructure for an installation should be designed to last at least for a decade. Typical ADP configurations change rapidly and require upgrades and replacement at least once in 3 years. New types of equipment find their way into the installation regularly. The infrastructure will thus have to cater to at least three generations of ADP equipment without major replacements or retrofits.

Different types of equipment behave differently and impose their own requirements on the infrastructure. Also, the mix of equipment is decided by the business involved. For example, a data center or an application service provider (ASP) business will consist of large number of high-capacity servers and few PCs. Typical end-user applications in any normal commercial business will most likely have predominantly desktop PCs and a few servers. Internet access providers will have a number of network switches and routers in addition to servers. The internal power supply devices may be of different designs and the facility designer must give a close look at the specifications of these devices.

The following aspects need consideration.

9.10.1 Power factor of loads

Most power supplies in ADP equipment are of switched mode type (SMPS) and unless corrected by suitably sized capacitors may draw currents at very low power factor, typically 0.6. For the same watt output, a device with lower power factor will draw a higher current. Accordingly, all internal and external wiring, transformers, circuit breakers and other power semiconducting devices in the connected circuits will have to be designed for higher ratings. This makes the design bulkier and more expensive.

Most high performance servers are nowadays provided with SMPS with power factor correction and draw load current at power factor close to unity. Since SMPS systems are specified in terms of watts, a lower PF will call for a higher kVA UPS (kVA = kW/PF).

It should also be remembered that the selection of battery from which the invertor gets its supply should take into consideration the active power handled. If the PF of the load is taken as 0.6 PF whereas it is actually 0.95, then a UPS system of a given kVA rating can be called upon to feed a larger kW load. It is easy to make such a mistake since the UPS systems are usually specified in terms of kVA rating. The battery selected may become undersized and may not work satisfactorily for the specified duration of backup time (usually 15 or 30 min).

9.10.2 Permissible voltage

Most SMPS loads can operate at voltages from 100 to 270 V. Thus, maintaining constant voltage is not always a necessity. This aspect needs attention since bypass source need not always have to be fed through a voltage stabilizer. The UPS voltage has to be matched with bypass source voltage at all times to ensure a smooth no-break change over.

9.10.3   Permissible frequency

Most ADP loads are insensitive to frequency. They can accept frequencies ranging from 47 to 63 Hz. This means that the UPS with higher-frequency window will be better. Such a design will permit the UPS to track the bypass supply through a larger range of frequency and maintain synchronization between UPS and bypass source. Such synchronization ensures that whenever a change over from UPS to bypass takes place, the same can be done without a break and therefore no danger of resetting of computer systems.

The UPS should have a high slew rate (ability for quick change in output frequency) so that the fall of frequency in the electrical system can be tracked without delay thus maintaining synchronization with bypass.

9.10.4   Backup time for batteries

The backup time for batteries should be selected so as to ensure that there is no need to shutdown systems under any condition including failure of the standby generator to start. Since the generators are usually started using auto mains failure (AMF) sensors, there may be a tendency to reduce the backup time by the designer. However, the failure of generator to start and need for operator intervention should be considered as a real possibility and backup time selected accordingly.

9.10.5   Harmonics

Since most computer loads draw currents with distorted waveforms, they impose a high degree of non-linearity in the system feeding them. Also, the harmonics from these devices can flow into the other loads of the system through capacitances (which offer a low impedance at higher frequency) and can cause unexplained heating of parts of the circuitry as well as erratic tripping of circuit breakers. Harmonics also cause the normal sinusoidal voltage waveform to get distorted thus making their effect felt throughout the rest of the system. A limit of 5% of harmonic current content must be aimed at it to avoid problems.

9.10.6   Standby generators feeding UPS loads

Generators feeding to ADP installations, which have large UPS equipment, will have to be designed taking into account the lower power factor and harmonic loading. Many generator manufacturers put limits on harmonic loading, which can be handled by the generators. This is due to the fact that harmonic voltages imposed by the non-linear loads can cause higher magnetic losses (due to the higher frequency) in all magnetic cores and active paths. This causes heating of the active parts and limits the generator from being run at their rated capacity. Also presence of capacitors in the load may cause resonance and the generator may fail to pick up voltage.

9.10.7   Short circuit behavior

It is necessary to design a UPS system in such a way that fault in one circuit fed by it does not cause total outage of the entire loads. Generally, the ability of the UPS to maintain an output voltage under short circuit conditions is limited since excessive currents will damage the semiconductor devices and the design of UPS has to incorporate current-limiting circuitry. This results in failure of circuit-protective devices to clear the fault since the energy flow is restricted to match the capacity of the semiconductors to handle. The resulting failure of UPS output voltage causes a change over to static bypass. The static bypass clears the short circuit by enabling adequate current flow and takes over the rest of the loads.

If the duration of voltage failure due to a short circuit (till the bypass restores it) is too long, it may cause the loads in the healthy circuit to reset also. To ensure this, the bypass must be in synchronism with the UPS at the instant of fault so that the change over takes place without a break. Hence the need to maintain large voltage and frequency window in the UPS design.

9.10.8 Crest factor

The UPS design should be done so that load inrush by switching of ADP equipment does not cause operation of the invertor protection. Mostly, this is not an issue since UPS systems have a crest factor of 3–5 whereas typical ADP requirements are limited to a crest factor of 2.

9.11 Grounding issues in static UPS configurations

As discussed earlier in this chapter, the issue of grounding UPS-derived systems is an important one as incorrectly grounded supply systems are unsafe; they result in equipment damage during faults/surges and also result in poor noise performance. We will discuss this issue in some detail.

In general, UPS configurations can be treated as separately derived source. A generator, transformer or convertor winding is a separately derived source if it has no direct electrical connection to supply conductors in another grounded system including a grounded circuit conductor. The UPS output is normally taken as a wye (star) connected winding galvanically isolated from the supply source feeding the UPS. However, many UPS systems are provided with a bypass circuit fed from the same source as the UPS without any isolation. The UPS cannot be considered as a separately derived source unless the bypass system has some form of galvanic isolation (such as a two-winding transformer).

The following are applicable in the case of separately derived sources:

- The neutral (grounded circuit conductor) should be bonded to the equipment safety grounding conductor and to the local ground such as building ground network or other made electrodes.
- The grounded conductor from neutral should be connected to the grounding conductor only at the source and not at any other point since it will make the ground fault protection ineffective.

We will now detail out the common UPS configurations and recommended grounding arrangements.

9.12 UPS configurations and recommended grounding practices

9.12.1 Configuration 1

In this case, the bypass source and the UPS source are the same and the bypass circuit has no isolation and is connected directly to UPS output. Thus, the definition of separately derived source is not satisfied. The UPS neutral is not therefore connected to the grounding conductor of equipment or to any local grounding electrode. Since there is no isolation between the source and the loads common-mode noise, attenuation is not ensured (Figure 9.18).

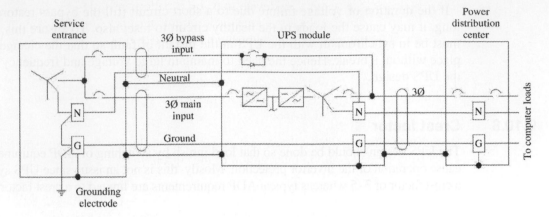

Figure 9.18
UPS system configuration 1

9.12.2   Configuration 2

In this case, the bypass supply is through a delta–wye transformer and thus there is galvanic isolation between the input supply to the UPS and the output under all conditions. The UPS can therefore be considered as a separately derived source. The neutral point of the UPS is bonded to the downstream equipment grounding wire as well as to the local grounding electrode. The bypass supply neutral is also bonded to UPS output neutral to provide a return path for neutral currents when bypass circuit is in operation (note that the static bypass switch is in-line wires only and the neutral connection is direct). Common-mode noise performance is better in this circuit if the neutral connection between bypass and UPS is kept short (Figure 9.19).

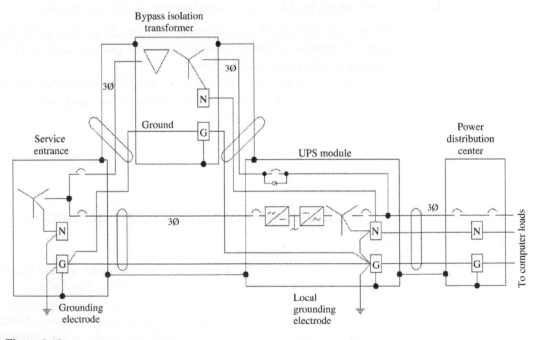

Figure 9.19
UPS system configuration 2

9.12.3   Configuration 3

This configuration shown in Figure 9.20 has a non-isolated bypass but the UPS output is taken to the loads through a distribution center, which incorporates an isolation transformer. Thus the UPS module is not a separately derived source by itself but the secondary of the isolation transformer is a separately derived source.

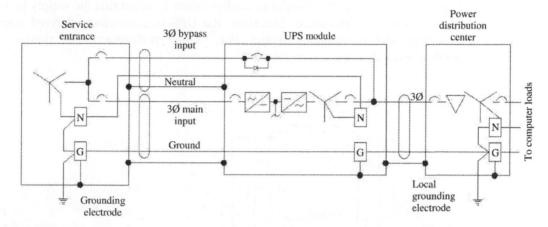

Figure 9.20
UPS system configuration 3

Thus the neutral of the UPS module is not bonded to the local bonding conductor. However, the isolation transformer neutral is bonded to the grounding wire from the computer loads fed by it as well as to the local grounding electrode system.

In this configuration, the power distribution center can be placed as close to the loads as possible so that it gives a better common-mode noise protection compared to the earlier configurations. The isolation transformer can also be used as a step down transformer permitting lower voltage supplies (208/220 V) to be served by UPS modules of higher voltage (380/415/480), which improves the cost-effectiveness of the design for UPS and wiring.

9.12.4   Configuration 4

This configuration (Figure 9.21) is similar to configuration 3 except that the service neutral is not brought to the UPS or the bypass module. Thus both UPS module and the distribution

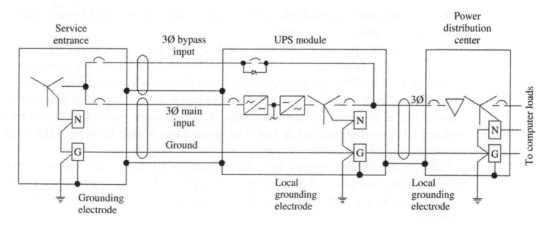

Figure 9.21
UPS system configuration 4

center can be treated as separately derived sources and neutral to ground connection is established in both these installations. Noise performance is similar to that of configuration 3.

9.12.5 Configuration 5

This case (Figure 9.22) is similar to configuration 2 except that the supply is from a delta-connected three-wire source. Therefore, the UPS is a separately derived source and the neutral/ground connections reflect this. Noise performance is similar to that of configuration 2.

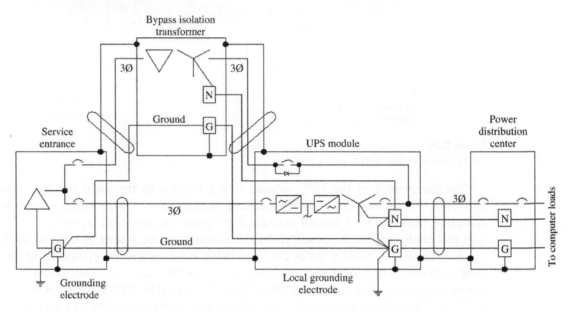

Figure 9.22
UPS system configuration 5

9.12.6 Configuration 6

This is an example of multiple UPS modules with an isolated bypass and a standalone static transfer switch configuration.

The combination of UPS modules and bypass can be treated as a single separately derived source. In order to create a common bonding point, the neutral of the bypass source and each UPS module are brought into the static switch module and connected to a neutral bar (Figure 9.23). A separate ground bus is also provided in the same cubicle. The neutral bar is bonded to the ground bus. All equipment grounding wires from the loads and the UPS are bonded to this bus. The ground bus is also bonded to the local grounding electrode and to the service ground.

This arrangement permits any UPS module to be taken out of service without affecting the integrity of grounding connections. Common-mode noise attenuation is also achieved by separation of the service neutral from the sensitive load supply neutral.

There are other possible UPS configurations, which we will not discuss here. But the general guiding principles in all these cases remain the same.

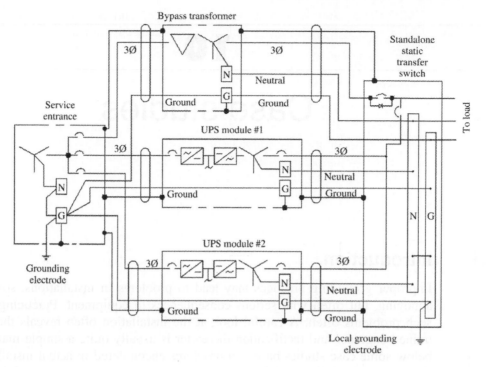

Figure 9.23
UPS system configuration 6

9.13   Summary

In this chapter, we covered the need for uninterrupted power sources and their role in improving power quality and reliability. We reviewed the basics of power quality and the definitions associated with this subject. We covered the methods of improving the quality of voltage. We saw about the ways in which uninterrupted power can be made available. We discussed about electromechanical and static equipment used for this purpose. We covered in detail the static UPS systems, these being the ones in common use, and their selection criteria when applied to ADP facilities. The issue of correct grounding practices necessary for various configurations of static UPS systems was also explained in detail. Unless correct grounding practices are adopted for UPS fed power distribution equipment appropriate to the configuration of the UPS, noise attenuation cannot be ensured and might result in malfunctioning of systems fed by them.

While this wraps up the basic course material, a few appendices have been added for providing further reading to the participants. The participants are advised to go through these in detail. Appendix F contains a set of exercises, which the participants will be required to solve, which can be followed by a discussion of the solutions by the course director. Answers are also provided for verification in Appendix G. In addition, a set of practical problems has been given under Appendix H for discussion among the group, as a group activity. These problems and the method adopted to tackle them are given in the form of case studies in the next chapter (Chapter 10). It is recommended, therefore to take up the last module (Chapter 10) for discussion after this group activity and the method adopted compared with the solutions arrived at by the group.

10

Case studies

10.1 Introduction

Improper grounding practices may lead to problems in installations, sometimes merely annoying, but often with serious consequences to equipment. Practicing engineers face such problems often. A careful look at the installation often reveals the problems to a trained engineer and rectification thereafter is usually quite a simple matter. We present below some case studies based on problems encountered in actual installations and how they were solved. Some of the participants might have come across such problems and we welcome them to share their experience with the group.

10.2 Case study 1

10.2.1 Effect of utility-induced surges

Problem

A steel mill with variable speed drives (VSDs) had problems of frequent tripping of the VSDs with the indication 'over-voltage in AC line'. Each tripping caused severe production disruption and resulted in considerable monetary loss due to lost production. Steady-state measurements by true RMS voltmeter showed that voltage was normal and within the specified operating range of the VDS. A power line monitor was then used in the distribution board feeding the VSDs and the incoming power feeder to the mill. At both locations, the monitors showed transient over-voltages of damped oscillatory type waveform with an initial amplitude of over 2.0 pu and a ringing frequency of about 700 Hz. The timing of disturbances coincided with the closing of capacitor banks in the utility substation feeding the steel mill (refer Figure 10.1a).

Analysis

It was confirmed by the VSD manufacturer that the VSDs were provided with over-voltage protection set to operate at 1.6 pu voltage for disturbances exceeding 40 μs. Since the switching transients were above this protection threshold, the VSDs tripped.

It may be noted that switching on a bank of capacitors results in high charging current inrush. When this current passes through the line's inductance L, a momentary voltage surge occurs. Further interaction of the capacitor C with inductance L results in an oscillatory flow of current, which is damped by the resistance R in the system. The oscillatory disturbance superimposed over the normal power-frequency voltage wave caused the over-voltage protection to operate.

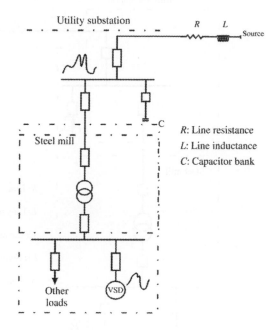

R: Line resistance
L: Line inductance
C: Capacitor bank

Figure 10.1a
Distribution arrangement

Solution

The solution lies in reducing the transient peak to a value that is below the over-voltage protection threshold. This was achieved in this case by installing a surge protection device (SPD) in each VSD. The SPD clamped the transient to a peak value of 1.5 pu thus avoiding the operation of over-voltage protection.

Another possible solution would have been to install an inductor $L1$ in the switching circuit of the capacitor for a few seconds and then shunt it by switch S. Since the voltage seen by the incoming feeder to the mill would be the combination across C and $L1$, the transient will have a smaller amplitude. This solution will however call for cooperation from the utility as it involves additional equipment to be installed by them (refer Figure 10.1b).

10.3   Case study 2

10.3.1   Effect of neutral breakage

Problem

In an office building with several offices, a constant voltage transformer (CVT) feeding a facsimile machine got overheated one night and started emitting smoke. On getting the fire-alarm signal, the security guard switched off supply to the machine. The following day, the engineer who was investigating this problem noticed that several fluorescent lamps going randomly on and off with an abnormal humming sound from the lamp chokes. This was happening all over the office intermittently.

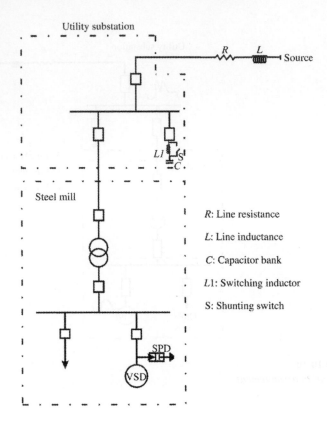

Figure 10.1b
After additions

Analysis

The office building was fed by a 500-kVA transformer, which fed a distribution center through a cable with three full capacity line conductors and a half capacity neutral conductor. It was found that in the incoming cable termination to the LV distribution center, the neutral lead was red hot and arcing intermittently when the loads were on.

Many offices in the building were installing computer systems fed by UPS equipment in the previous 1 year. The non-linear nature of these loads had caused much higher neutral currents to flow compared to a balanced three-phase load, which the cable with reduced neutral was designed to carry. This had caused the neutral termination to overheat and deteriorate causing intermittent loss of neutral continuity (refer Figure 10.2).

When the neutral connection was absent, the voltage across each phase group was decided by the load distribution between the three phases and most loads being of single-phase type loads were not perfectly balanced. Refer to the impedance and phasor diagrams in Figure 10.2.

The heavily loaded phase was thus having a smaller voltage and as the loads were being switched on in the morning hours, it had caused random voltage variation in different phases. This in turn resulted in flickering of the fluorescent loads and humming of chokes in the phases having higher voltage (lightly loaded phases). The CVT had burnt out due to prolonged over voltage.

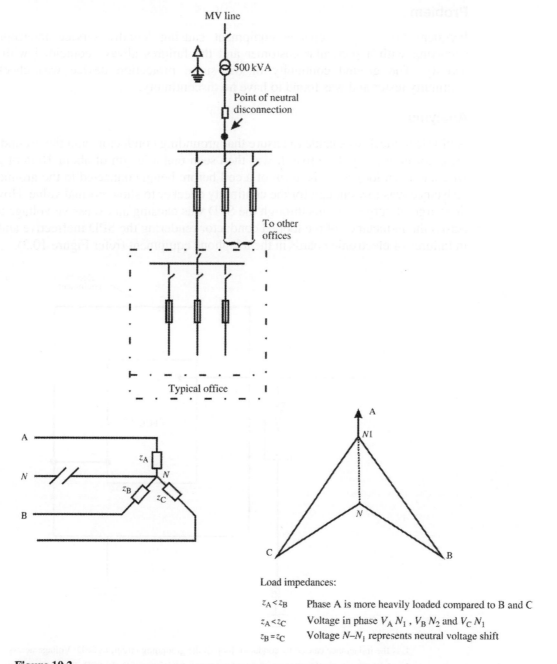

Load impedances:

$z_A < z_B$	Phase A is more heavily loaded compared to B and C
$z_A < z_C$	Voltage in phase $V_A N_1$, $V_B N_2$ and $V_C N_1$
$z_B = z_C$	Voltage $N-N_1$ represents neutral voltage shift

Figure 10.2
Distribution arrangement, impedance and phasor diagrams

Rectification

As an immediate step, the termination was repaired and the cable put back in service. As a preventive measure, the cable was replaced by one having a neutral conductor of the same size as the phase conductors. The system has been operating now for about 5 years without any failures.

10.4 Case study 3

10.4.1 Telephone equipment failures

Problem

Frequent failure of telephone equipment causing lengthy service interruptions was occurring with a particular customer and the failures always coincided with lightning activity. The ground continuity of the surge protection device was checked by a continuity tester and was found to have no discontinuity.

Analysis

A physical check was made to ensure that grounding conductors and the ground electrode were not in anyway defective. It was then seen that a length of about 10 m of grounding wire had been looped in the form of a coil before being connected to the ground rod. The resistance was low enough for the continuity checker to show normal value. However, the discharge of surge currents through the SPD was causing an excessive voltage to develop across the inductance of the looped conductor rendering the SPD ineffective and resulting in failures of electronic boards in the telephone equipment (refer Figure 10.3).

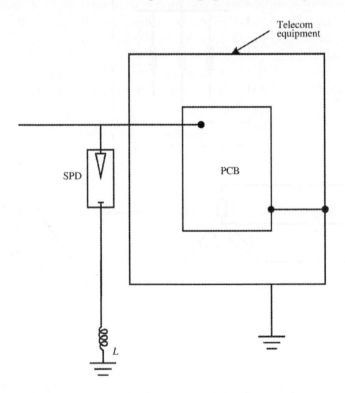

L is the inductance caused by conductor loop in the grounding circuit of SPD. Voltage across

L = L·dI dT, where dI is the rate of rise of discharge current through the SPD. PCB fails to this additional voltage impressed between the active circuits of the PCB and its frounding bus.

Figure 10.3
Connection of communication equipment

Rectification

The grounding conductor was cut and re-terminated with the shortest possible path for the discharge current flowing from SPD to the ground electrode. There were no further failures.

10.5   Case study 4

10.5.1   Effect of isolated grounding

Problem

In a large office complex, computer errors including system crashes/reboots were happening during thunderstorms. The grounding of computer system had been carried out according to the recommendations of manufacturers. The grounding leads were insulated and terminated to an isolated grounding bus. This bus was connected to a grounding electrode consisting of multiple driven rods well away from the building.

Analysis

It was suspected that the isolated computer grounding system was responsible for this problem. Measurements were made using a power quality monitor to confirm this. It was noticed that appreciable voltages were recorded between the building grounding system and the computer ground during electrical storms. This presented a safety hazard to personnel using these computer systems and was also introducing noise into the systems through the capacitances between the computer systems and the building ground (refer Figure 10.4).

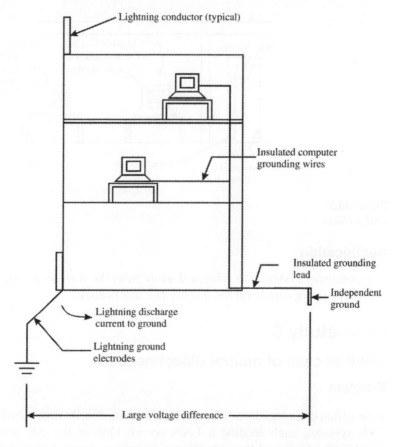

Note: Lightning current flowing through the lightning conductor to ground electrodes causes the potential of the building members to rise above that of the causing system errors. Unsafe potentials can also arise.

Figure 10.4
Isolated grounding

Rectification

The manufacturer of the computer systems was consulted and the computer ground system and the building ground were bonded together on a trial basis. It was noticed that the system problems were no longer occurring. The grounding wires of computer systems were also routed along with the power supply wires to the systems for better performance.

10.6   Case study 5

10.6.1   Problem of monitor display

Problem of wavy monitor display was reported from a new computer installation. Power quality checks did not indicate any abnormality. Subsequently, magnetic field measurements were carried out and indicated presence of high power frequency magnetic fields.

Analysis

When the surrounding area was inspected, it was found that on the other side of the wall where the computer was located, a major power distribution center was located. The high magnetic fields caused by the currents were causing the problem with the monitor display (refer Figure 10.5).

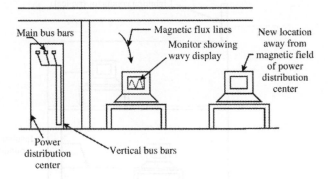

Figure 10.5
EMI problem

Rectification

The computer system was relocated away from the wall by a distance of about 1.5 m. The waviness disappeared and the display became normal.

10.7   Case study 6

10.7.1   Another case of neutral disconnection

Problem

In an office computer installation, a single-phase distribution board was feeding two 1 kVA UPS systems, each feeding a LAN server. One of the AC units whose power feeder developed a fault was temporarily connected to a spare outlet of the DB feeding the UPS systems. All of a sudden, the circuit breakers on the AC power input of both UPS systems tripped. Power to the AC unit was also interrupted though the feeder to the unit was ON. Switching on the UPS, incoming CBs restored normalcy but the tripping happened again within a few seconds.

Analysis

Voltages were measured between both phase and neutral and between neutral and ground. The voltages were normal when UPS units were functioning (AC unit was OFF). Currents were measured by a clip-on ammeter. The neutral circuit of the UPS did not indicate any current. When the AC unit was switched on, both neutral circuits indicated higher than normal currents before the circuit breaker interrupted the current.

Further checking showed that the neutral connection at the incoming line to the DB was bad. Also, within the UPS an inadvertent connection between neutral and ground was found (refer Figure 10.6a).

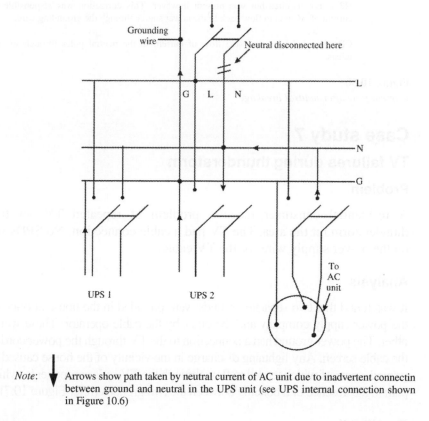

Note: Arrows show path taken by neutral current of AC unit due to inadvertent connectin between ground and neutral in the UPS unit (see UPS internal connection shown in Figure 10.6)

Figure 10.6a
Distribution arrangement

As long as only UPS units had been connected, the problem with broken neutral connection was not evident since the return current was flowing through the grounding wire of the respective UPS system. When the AC unit was connected, its neutral current took a path through the neutral pole of the circuit breakers of the UPS AC power input (dividing between both UPS breakers). The breakers tripped due to the starting current inrush of the compressor motor of the AC unit (refer Figure 10.6b).

Rectification

The neutral connection was restored. The connection between ground and neutral of the UPS systems were removed. The AC unit could be run without any problem till its own supply was repaired. A residual current CB was added in the incoming side of the DB to be able to pinpoint such problems as soon as they occur.

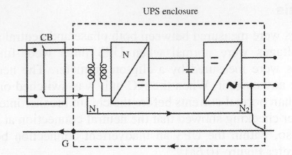

N2 was grounded through G as the Ups output is a separately derived source. Correction N1-N2 is not required but was present however. This correction was responsible for neutral current of AC unit to flow back to the newer source through the grounding wire.

CB tripped due to excessive flow of current in the neutral poles through its over-current release.

Figure 10.6b
Current flow after neutral breakage

10.8   Case study 7

10.8.1   TV failures during thunderstorm

Problem

A residential consumer faced a problem of repeated TV set failures whenever a thunderstorm hit the area. The TV had a cable connection. No SPDs were provided either on the power supply wires or the TV cable.

Analysis

It was found that two separate grounds were provided in the house at opposite ends. One was by the power supply company and the other by the cable operator. These were not bonded to each other. The power ground had a connection to the TV through the power cord and the other through the cable screen. Any lightning discharge in the vicinity of the house caused flow of current in the ground, some of which tended to flow through the TV set's components, which provided a parallel path in the absence of a bonding between the two grounds (refer Figure 10.7).

Rectification

Approved surge protection devices were provided for both power and cable TV circuits. The ground electrodes were bonded together by an adequately sized copper wire. The problem of TV failure during thunderstorms ceased thereafter.

10.9   Case study 8

10.9.1   Potential difference between buildings

Problem

In an industrial facility, there was a cluster of four buildings each having a process control computer connected with data cabling. Each computer was grounded to the grounding system of the respective building. The grounding systems of the buildings were interconnected through water mains and metallic sheaths of cables, etc. The functioning of the computer systems was very erratic (refer Figure 10.8a).

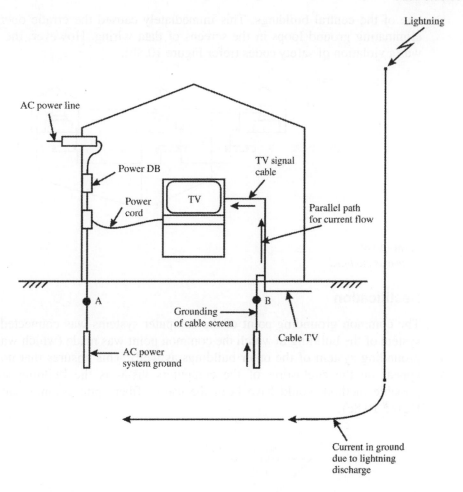

To avoid destructive current flow through TV points A and B need to be connected through a bonding wire.

Figure 10.7
Failure of TV set due to multiple (unbounded) grounds

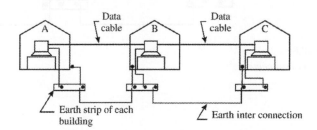

Figure 10.8a
System as existing

Analysis

It was thought that the possible reason for erratic operation was the flow of stray ground currents through the earthed screens of the data wiring. The grounding of the computer installations of the four buildings was brought together and connected at a single point in one of the central buildings. This immediately caused the erratic operation to cease by eliminating ground loops in the screens of data wiring. However, the new arrangement was a violation of safety codes (refer Figure 10.8b).

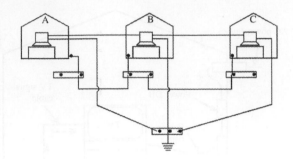

Figure 10.8b
System as modified

Rectification

The common grounding point of the computer systems was connected to the grounding system of the building in which the common point was made (which was connected to the grounding system of the other buildings as well). This ensures that no unsafe potentials appear on the enclosures of the computers vis-à-vis the building structures. Another possible method would have been the use of fiber-optic communication cabling (refer Figure 10.8c).

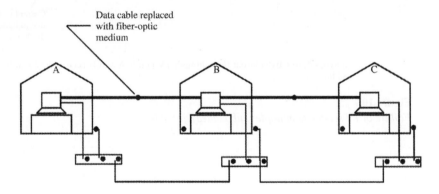

Figure 10.8c
Preferred arrangement

Appendix A

Grounding regulations from various national codes

A.1 USA – National Electrical Code (NEC) regulations relating to grounding practices

The design, installation and maintenance of grounding connections in electrical systems have been exhaustively studied and documented. Mandatory regulations have been established and are in use by different countries (example: NEC/NFPA codes of USA. Article 250 of NEC code of USA is dedicated to the subject of grounding). These regulations usually stipulate the minimum requirements to ensure safety of equipment and personnel but system designers are free to improve on them on the conservative side.

The NEC defines service equipment as switching and protective equipment installed at the point of entry of power from the electric power utility to the consumer premises. The provisions of NEC are meant to ensure that the electrical installation on the downstream side of the service equipment is free from defects that can cause fire, explosion or electric shock hazards under normal or fault conditions of the system. NEC recommends that all grounding electrodes including cold-water piping, metallic building frames, etc. used as grounding electrodes are bonded to the equipment grounding system at the service equipment ground.

As discussed in the chapter earlier, bonding of all building grounding in this manner ensures that no dangerous potential develops between the equipment grounding conductors, metal raceways, building structures, cold-water mains, etc. even though their potential with reference to the mean earth potential may be substantially higher. Any voltage difference between these points will be due to abnormal conditions such as a ground fault. But even so, the potential difference will not reach the touch potential limits thus ensuring that the system is free of shock hazards.

Section 250-118 of NEC: 1999 permits the use of various grounding conductors including tubular conduits used as cable raceway, cable armor and cable trays. Where grounding conductors are employed, it must be ensured that they are sized to withstand ground fault currents of value and duration appropriate to the circuit under consideration.

If the conductor sizing is inadequate, the following may result:

- *Damage to the insulation of the ground conductor:* In case the conductor is bare, insulation damage may occur in the phase conductor with which it is in contact.

- *Fusing of the ground conductor in extreme cases:*   Though this will result in the interruption of the fault current, it will cause the potential of the enclosure of the faulted equipment to rise beyond acceptable limits.

The maximum temperature of the conductor during a fault can be calculated by the formula:

$$I^2 t/A = 0.0297 \; (\log \; (T_m + 234)/(T_i + 234))$$

Where

I is the Fault current through the conductor is Amps

A is the conductor cross sectional area in circular miles

t is the time of fault in seconds

T_i is the intial operating temperature in ° C

T_m is the maximum temperature permissible in ° C

For the purpose of reviewing the selection, the following value of T_m can be used:

- Copper fusing temperature   800 °C.

Maximum permissible temperature of insulation:

- For thermo-plastic insulation   150 °C
- For XLPE insulation   250 °C.

Current and time setting of interrupting device should not cause unacceptable temperature rise of the grounding conductor. If, however, this exceeds the limits of temperature, the following are the options:

- Increase the size of the grounding conductor
- Adding special ground fault equipment to the system to sense even low value of earth fault currents and trip the circuit faster.

A.2   South Africa – grounding practices as per SABS standards

A.2.1   Grounding of remote electrical installations fed from pole-mounted transformers

The grounding requirements associated with remote MV/LV transformer-fed installations are somewhat different from the approach adopted normally in industrial installations.

Industrial power systems typically use the scheme shown in Figure A.1. It may be noted that the transformers are cable fed and they do not incorporate surge arrestors. The transformer LV neutral is grounded directly to an earth electrode and at the same time connected to the plant grounding network. The tank is connected to the ground network using at least two earth leads. The LV system neutral and the ground network of the plant to which the safety ground of all equipment is connected are thus bonded all through with metallic connections and faults to enclosures (as shown and explained in Chapter 3, Figure 3.5).

Note: TN-S refers to classification explained later in this appendix.

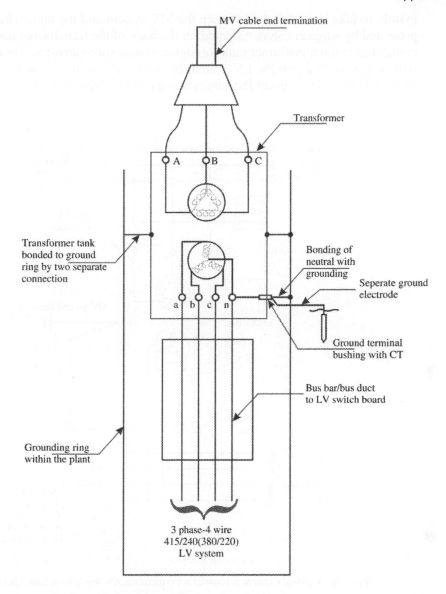

Note: This arrangement is followed in TN-S systems where transformer neutral point is connected to ground only at three points and no where else.

Figure A.1
Typical industrial power transformer installation

On the other hand, the situation in a remotely installed pole-mounted transformer fed by an overhead line and again feeding isolated consumers through LV overhead lines is somewhat different. Also, the LV installations follow the TN-CS/PME type of connections as explained later in this appendix. Moreover, the ground electrode resistance cannot be expected to be too low either.

The arrangement suggested in SABS standards for this type of distribution is shown in Figure A.2.

It is noteworthy that the installation recommends the LV neutral to be grounded at a point well away from the transformer (usually at the first LV pole). The tank earthing is mainly to take care of faults between the MV system and the tank. The MV windings are protected by surge arrestors mounted on the body of the transformer and the LV neutral is connected to the transformer tank through a neutral surge arrestor. Though the MV surges will be transmitted into the LV system by the coupling effect, the high surge impedance of the LV lines will prevent the surges being propagated into the LV system.

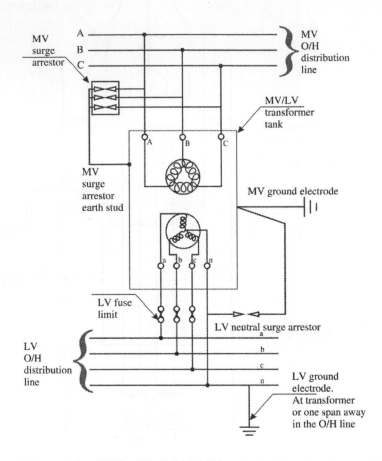

Note: If LV ground electrode is installed at the transformer, it should be at least 5 m away from MV electrode.

Figure A.2
Grounding of remote distribution transformers/LV system

Also the surges being of very short duration will not pose a safety hazard. This design takes into account the following factors:

- Need to prevent the LV neutral from assuming dangerous touch potential when an MV to tank fault occurs
- Need for limiting the voltage across the neutral arrestor
- Need for detecting MV to tank faults by MV earth fault protection.

The detailed analysis of the design approach is given in the next few paragraphs. The maximum electrode resistance permissible under these conditions is also arrived at. The ground electrode resistance values thus arrived at are used to arrive at standard ground electrode configurations described in greater detail in this appendix.

A.2.2   Analysis of typical pole-mounted MV/LV transformer grounding

We will now discuss in some detail the reasons for the grounding philosophy adopted for pole-mounted MV/LV transformer substations illustrated in Figure A.2.

LV grounding

The purpose of LV system grounding is as follows:

- To maintain the LV neutral potential to as close to earth potential as possible thereby prospective touch voltages in all the grounded metal parts of equipment
- To provide a low-impedance return path for any LV ground faults
- To ensure operation of MV protection in the event of an inter-winding fault (MV and LV) within the transformer.

MV grounding

The surge arrestors of MV lines are connected to the transformer tank, which in turn is grounded through the MV, ground electrode. This limits the voltage between the tank and the lines to the voltage drop across the arrestors in the event of a surge. In case the arrestors are connected by a separate lead to the ground, the voltage drop across the resistor would also additionally appear between lines and tank and cause insulation failure.

Combined MV/LV grounding

Though it is theoretically possible to have a combined ground at the transformer for both MV and LV, such a practice may lead to unsafe conditions in the event of an MV to LV fault. Figure A.3 shows the reason.

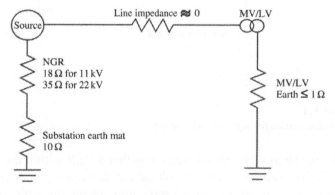

Figure A.3
Equivalent circuit for combined MV/LV grounding

The total impedance for a fault between HV and MV winding (neglecting the line impedance and the leakage impedance of the transformer windings) is the substation ground mat resistance of 10 Ω, the NGR (neutral grounding resistance) value and the MV/LV combined ground electrode resistance assumed as 1 Ω.

For a 22-kV system (with line to ground voltage of 12 700 V) the current flow is:

$$I_G = \frac{12\,700}{(10 + 35 + 1)}$$

35 Ω being the NGR value for a 22-kV system.

This gives a figure of 276 A. This current will cause the potential of 276 V to appear on the transformer tank and through the neutral lead to the enclosures of all equipment connected in the LV system with respect to true earth potential (this is because in TN-C-S type of systems, which we saw in the last chapter, the neutral and equipment ground are one and the same). This value is unacceptably high. In actual practice, the line and transformer impedances come into play and the value will therefore get restricted to safe values. Use of combined MV and LV grounding is therefore possible only if the ground resistance can be maintained below 1 Ω.

Separate MV/LV grounding

MV electrode

In view of the difficulty of maintaining a very low combined ground resistance arrived at above, the code allows the use of a separate ground for the LV neutral away from the transformer. The only point of connection between the LV system and the transformer tank is the LV neutral surge arrestor whose grounding lead is connected to the transformer tank (refer Figure A.4).

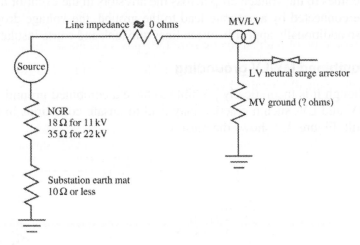

Figure A.4
Equivalent circuit for separate MV ground

The problem with this connection is that a fault within the transformer (MV winding to core fault) resulting in rise of voltage can cause a high enough voltage to ground causing the neutral surge arrestor to fail and communicate the high voltage into the LV system.

Assuming a maximum LV voltage of 5000 V for withstand of neural surge arrestors, the voltage rise across the MV ground electrode resistance should not be greater than this value. For a 22-kV system (with line to ground voltage of 12 700 V) the ground electrode voltage can be calculated using the potential division principle as follows.

$$\frac{5000}{12\,700} = \frac{Rm}{Rm + 10 + 35}$$

where *Rm* is the resistance of MV ground electrode. It can be calculated that Rm can have a value of 29 Ω to be able to limit the voltage.

For 11-kV system, a value of 100 Ω is permissible. The limit for the electrode resistance should also consider the ground fault current so that the MV ground fault relay can operate reliably to isolate the fault.

A value of 30 Ω is taken as the limit for ground electrode systems for all MV systems. Standard configurations are available in the code for 30 Ω electrodes (refer Chapter 6) and can be used in the design.

LV electrode

The LV electrode resistance should be normally expected to permit sufficient fault currents for detection. Since with the LV line to neutral voltage of 240 V, the resistance limit works out to 2.4 Ω if a ground fault current of 100 A is to be obtained. However, with the TN-C-S type of system, all equipment enclosures are directly connected to the neutral at the service inlet itself and thus the current flow does not involve the ground path at all.

So, the limit of LV grounding resistance is decided by the criteria of obtaining sufficient fault current when there is an MV to LV fault without involving the tank or core (refer Figure A.5).

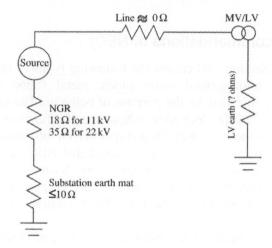

Figure A.5
Equivalent circuit for separate LV ground

Assuming an MV earth fault protection setting of 40 A, the ground loop resistance can be arrived at 318 Ω (12 700/40) for 22-kV system. The permissible ground electrode resistance works out to 273 Ω (after taking off the values of NGR and substation ground resistance). If we consider a safety factor of 400%, the maximum value of LV ground resistance can be taken as 68 Ω. The safety factor will ensure that the seasonal changes of soil resistivity will have no adverse effect on protection operation. Standard configurations are available in the code for 70 Ω electrodes (refer Chapter 6) and can be used in the design.

A.3 Ground electrode recommendations by different national codes

Examples are given from the Australian Standards (AS 3007.2), National Electric Code (USA) and the South African Standards below to illustrate three different approaches taken for the design of ground electrodes.

A.3.1 Australia – New Zealand (Standard AS 3007.2)

The following types of ground electrodes are permitted by the standard:

- Rod(s) or pipe(s)
- Tape(s) or wire(s)

- Plate(s) or mat(s)
- Electrode(s) embedded in foundations
- Metallic reinforcement of concrete
- Other suitable underground structures.

One or more ground electrodes suitable for the soil conditions and value of earth resistance required can be selected and used in combination if required.

The material and total cross-sectional area of the ground electrode(s) shall be so as to provide a conductance of not less than that of the grounding conductor required for the application.

Metallic pipe systems for water or other services (for example, flammable liquids or gases, heating systems) shall not be used as ground electrodes for protective or functional purposes.

A.3.2   USA (recommendations of NEC)

The NEC Section 250 covers the following types of electrodes. Those in the first type are metal underground water pipes, metal frame of building, etc. that are not specifically installed for the purpose of being used as ground electrodes. The other type is known as 'made electrodes' whose specific function is for use as ground electrodes. Examples of these are metal rods/pipes driven into ground, concrete-encased electrodes, buried ground ring, etc. It is to be noted that NEC does not require any special ground electrodes in an installation where some form of buried underground metalwork is already present and can be used for this purpose (this aspect of NEC is totally at variance with the prescription of the Australian Standard mentioned in the earlier paragraph).

Metal underground water mains used as ground electrodes should be in direct contact with ground for a minimum distance of 10 ft and made continuous by bonding around insulating joints. Also, continuity should not depend on water meters, filters or any other removable devices.

An effectively grounded metal frame or structure of a building can also serve as a ground electrode. An electrode encased in two inches of concrete located within a concrete foundation footing near the bottom part, which is in direct contact with the ground, can be used as a grounding electrode.

A buried ground ring around a building made of at least 20 ft of copper conductor at a depth of over 2.5 ft can also be used as an electrode. Underground metal gas piping system shall not be used as an electrode. Also NEC code does not permit the use of aluminum as suitable material for ground electrodes.

Where electrodes of above types are not readily available, electrodes specifically made for use as ground electrodes will have to be deployed. These electrodes shall be driven into the soil in such a way that a substantial part of their length is below the permanent moisture level of the ground in the area under consideration. When more than one electrode is used they shall be separated by more than 6 ft and preferably by one-rod length and shall be effectively bonded so as to result in a single composite electrode system. The length of electrodes shall not be lower than 8 ft and in the case of pipes/conduits the diameter shall not be less than ¼ in. trade size. Rods shall not be lower than 5/8 in. diameter for use as made-electrodes. The electrodes shall be galvanized or otherwise metal coated to minimize corrosive effects of being buried in the ground. Plate electrodes of 2 sq. ft surface area and ¼ in. thick made of iron/steel plates and buried in soil at a depth of 2.5 ft can also be used as electrodes.

All non-conductive coatings such as paint/enamel and rust shall be scraped of the surface of the electrodes to ensure proper contact with soil.

Resistance of made-electrodes shall not be lower than 25 Ω. In case this is not so, multiple electrodes bonded together shall be used to bring down the resistance value.

A.3.3 South Africa (Standard SCSASAAL9)

This standard recommends ground electrodes using buried horizontal conductors, vertically driven electrodes or a combination of both. Buried horizontal conductors are easier to install and result in less steeper potential gradients during a ground discharge. This type of electrode (called as trench electrode) must be installed as deep as practicable but not less than 500 mm below ground level.

The reasons for this are:

- Less prone to mechanical damage
- Better current dissipation using the layer of soil above the conductor
- Lower voltage gradient during discharge.

Vertically installed electrodes may be extended to substantial lengths (15–90 m depth) and are able to make contact with low-resistivity soils. At these depths the soil resistivity is same throughout the year and less prone to seasonal variations that are found on the soil layers near the surface. Vertical rods are also found to have superior surge performance and are useful for electrodes meant for lightning protection of structures.

Electrodes constructed from a combination of driven rods and radial array of buried conductors ensure a stable resistance value and good performance for power frequency and high-frequency ground discharges. They minimize the chance of high soil potential gradients and avoid failure of ground connection due to a single electrode conductor breakage.

This standard specifies the following type of electrodes:

- Non-extendable ground rods in a multi-electrode (preferably 3-point star) configuration
- Linear trench electrode
- Extendable ground rods
- Vertically installed conductor.

Linear trench electrodes are recommended where vertical rods are not convenient to use.

This standard defines the minimum resistance of the ground electrode in LV installations neutral as 68 Ω and that of the transformer tank ground electrode (in MV installations of 11 or 22 kV) as 30 Ω. The reasoning behind this stipulation was explained in Section A.2.

Standard SCSASAAL9 defines various electrode configurations having resistance of 30, 70 and 150 Ω in different soils in the form of tables for easy selection.

Recommended arrangement of electrodes

Figure A.6 shows various possible arrangements of electrodes referred in the tables for electrode configurations.

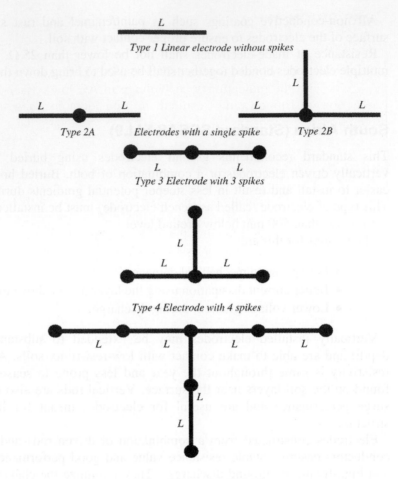

Figure A.6
Arrangements of ground electrodes

Tables in Figures A.7 to A.9 show the electrode configurations required for achieving different target values of ground electrode resistance.

Electrode Type	Dimension L in meters for Soil Resistivity (Ω m)			
	300	**600**	**900**	**1500**
Type 1	19	41	65	120
Type 3	8	20	32	58
Type 4	5	**	**	**
Type 5	**	13	22	40

Figure A.7
Electrode configuration for 30-Ω resistance

Electrode Type	Dimension L in meters for Soil Resistivity (Ω m)			
	300	600	900	1500
Type 1	7	16	26	46
Type 2A	3	**	**	**
Type 2B	2	**	**	**
Type 3	**	6.5	12	22
Type 4	**	4	8	15

Figure A.8
Electrode configuration for 70-Ω resistance

Electrode Type	Dimension L in meters for Soil Resistivity (Ω m)			
	300	600	900	1500
Type 2B	1	2	3.5	6.5

Figure A.9
Electrode configuration for 150-Ω resistance

Notes:

- Figures in the tables shown in Figures A.7 to A.9 have been arrived based on the assumption that all horizontal conductors are buried at a depth of not less than 0.5 m and the spikes (vertical) are 1.5-m long.
- In case the soil resistivity is between two of the values, the configuration for the higher value must be chosen.

Table shown in Figure A.10 can be used for recording the ground resistivity test values and calculate the ground (soil) resistivity.

Spacing Between Probes S in meters	Specific Depth $D = 0.8 \times S$	Test Reading R Ohms	Factor $K = 2 \times P \times S$	Resistivity ($R \times K$)
1	0.8		6.28	
2	1.6		12.57	
3	2.4		18.85	
5	4		31.42	
10	8		62.83	
15	12		94.25	

Figure A.10
Ground resistivity measurement

A.4 Measurement of ground electrode resistance – recommended practice by South African standard SCSASAAL9

Three different methods of resistance measurement have been suggested in the above standard for use with ground electrodes. The simplest method known as '61.8% method' is used for relatively small ground electrode systems, which are symmetrical about a point (the electrical center). Current injection electrodes need to be placed at long distances from the electrical center, which could be a disadvantage.

The measurements made in this method have to be verified using 'Four potential method'. This requires the current probes to be located at large distances from the ground electrode.

The third method called 'the slope method' does not require knowledge of the electrode center and can also be carried out using current probes at relatively smaller distance from the electrode. This method is also more accurate.

A.4.1   61.8% Method of measurement

The arrangement is shown in Figure A.11. In this method, the outer (current) leads are connected to the ground electrode whose resistance is to be measured and a spike placed at a distance of five times the longest diagonal of the ground electrode system. The distance should not be less than 50 m and is preferably standardized as 100 m. One potential lead is connected to the electrode under test and the other at a distance of 61.8% of the distance between current leads as shown in the figure. The resistance value obtained by the instrument is the electrode resistance.

A.4.2   Four potential method of measurement

The arrangement is shown in Figure A.12. This test setup is similar to the previous method except that the second potential probe is kept at different distances between the two current leads and readings taken. Values measured at 20, 40, 50, 60, 70 and 80% of the distance between the current leads are designated as $R1$, $R2$, $R3$, $R4$, $R5$ and $R6$.

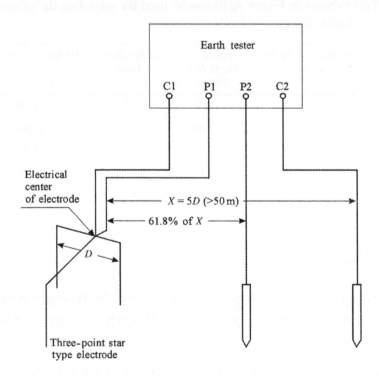

Figure A.11
Measurement arrangement for 61.8% method

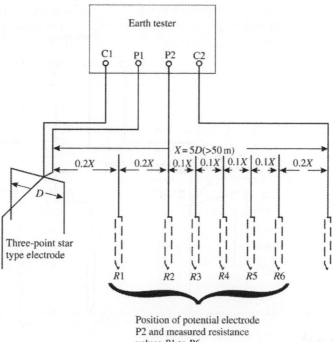

Figure A.12
Measurement arrangement for four potential method

The electrode resistance R can be calculated using the following different equations:

$$R = -0.1187 \times R1 - 0.4667 \times R2 + 1.9816 \times R4 - 0.3961 \times R6$$
$$R = -2.6108 \times R2 + 4.0508 \times R3 - 0.1626 \times R4 - 0.2774 \times R6$$
$$R = -1.8871 \times R2 + 1.1148 \times R3 + 3.6837 \times R4 - 1.9114 \times R5$$
$$R = -6.5225 \times R3 + 13.6816 \times R4 - 6.8803 \times R5 + 0.7210 \times R6$$

The values should be nearly equal. In case the first value is very much different from the others, it can be ignored. The average of the acceptable values can be taken as the electrode resistance.

This can be taken as a check value for the measurement done by 61.8% method of measurement.

A.4.3   Slope method of measurement

The arrangement is shown in Figure A.13. The first current lead is connected to any point on the ground electrode and the second current lead at a distance of more than 50 m from the first. The first potential lead is connected to the ground electrode and the second kept at distances of 20, 40 and 60% of the distance between current electrodes and resistance measured. These values are designated as $R1$, $R2$ and $R3$, respectively.

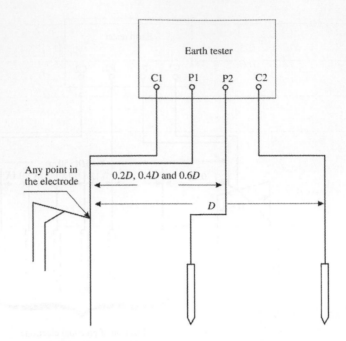

Figure A.13
Measurement arrangement for slope method

The slope coefficient $M = (R3 - R2)/(R2 - R1)$ to 3 decimal accuracy.

Using the table given in Figure A.14, the distance ratio R can be found out. The first column is for the first two decimal places of M. The value of R can be obtained using the

M	0	1	2	3	4	5	6	7	8	9
0.40	0.6432	0.6431	0.6429	0.6428	0.6426	0.6425	0.6423	0.6422	0.6420	0.6419
0.41	0.6418	0.6416	0.6415	0.6413	0.6412	0.6410	0.6409	0.6408	0.6406	0.6405
0.42	0.6403	0.6402	0.6400	0.6399	0.6437	0.6396	0.6395	0.6393	0.6392	0.6390
0.43	0.6389	0.6387	0.6386	0.6384	0.6383	0.6382	0.6380	0.6379	0.6377	0.6376
0.44	0.6374	0.6373	0.6372	0.6370	0.6369	0.6367	0.6366	0.6364	0.6363	0.6361
0.45	0.6360	0.6359	0.6357	0.6356	0.6354	0.6353	0.6351	0.6350	0.6348	0.6347
0.46	0.6346	0.6344	0.6443	0.6341	0.6340	0.6338	0.6337	0.6336	0.6334	0.6333
0.47	0.6331	0.6330	0.6328	0.6327	0.6325	0.6324	0.6323	0.6321	0.6320	0.6318
0.48	0.6317	0.6315	0.6314	0.6312	0.6311	0.6310	0.6308	0.6307	0.6305	0.6304
0.49	0.6302	0.6301	0.6300	0.6298	0.6297	0.6295	0.6294	0.6292	0.6291	0.6289
0.50	0.6288	0.6286	0.6285	0.6283	0.6282	0.6280	0.6279	0.6277	0.6276	0.6274
0.51	0.6273	0.6271	0.6270	0.6268	0.6267	0.6265	0.6264	0.6262	0.6261	0.6259
0.52	0.6258	0.6256	0.6255	0.6253	0.6252	0.6252	0.6248	0.6247	0.6245	0.6244
0.53	0.6242	0.6241	0.6239	0.6238	0.6236	0.6235	0.6233	0.6232	0.6230	0.6229
0.54	0.6227	0.6226	0.6224	0.6223	0.6221	0.6220	0.6218	0.6217	0.6215	0.6214
0.55	0.6212	0.6210	0.6209	0.6207	0.6206	0.6204	0.6203	0.6201	0.6200	0.6198
0.56	0.6197	0.6195	0.6194	0.6192	0.6191	0.6189	0.6188	0.6186	0.6185	0.6183
0.57	0.6182	0.6180	0.6179	0.6177	0.6176	0.6174	0.6172	0.6171	0.6169	0.6168
0.58	0.6166	0.6165	0.6163	0.6162	0.6160	0.6159	0.6157	0.6156	0.6154	0.6153
0.59	0.6151	0.6150	0.6148	0.6147	0.6145	0.6144	0.6142	0.6141	0.6139	0.6138
0.60	0.6136	0.6134	0.6133	0.6131	0.6130	0.6128	0.6126	0.6125	0.6123	0.6121

M	0	1	2	3	4	5	6	7	8	9
0.61	0.6120	0.6118	0.6117	0.6115	0.6113	0.6112	0.6110	0.6108	0.6107	0.6105
0.62	0.6104	0.6102	0.6100	0.6099	0.6097	0.6096	0.6094	0.6092	0.6091	0.6089
0.63	0.6087	0.6086	0.6084	0.6083	0.6081	0.6079	0.6076	0.6076	0.6074	0.6073
0.64	0.6071	0.6070	0.6068	0.6066	0.6065	0.6063	0.6061	0.6060	0.6058	0.6057
0.65	0.6055	0.6053	0.6052	0.6050	0.6049	0.6047	0.6045	0.6044	0.6042	0.6040
0.66	0.6039	0.6037	0.6036	0.6034	0.6032	0.6031	0.6029	0.6027	0.6026	0.6024
0.67	0.6023	0.6021	0.6019	0.6018	0.6016	0.6015	0.6013	0.6011	0.6010	0.6008
0.68	0.6006	0.6005	0.6003	0.6002	0.6000	0.5998	0.5997	0.5995	0.5993	0.5992
0.69	0.5990	0.5989	0.5987	0.5985	0.5984	0.5982	0.5980	0.5979	0.5977	0.5976
0.70	0.5974	0.5973	0.5971	0.5969	0.5967	0.5965	0.5964	0.5962	0.5960	0.5959
0.71	0.5957	0.5955	0.5953	0.5952	0.5950	0.5948	0.5947	0.5945	0.5943	0.5942
0.72	0.5940	0.5938	0.5936	0.5935	0.5933	0.5931	0.5930	0.5928	0.5926	0.5924
0.73	0.5923	0.5921	0.5920	0.5918	0.5916	0.5914	0.5912	0.5911	0.5909	0.5907
0.74	0.5906	0.5904	0.5902	0.5900	0.5899	0.5897	0.5895	0.5894	0.5892	0.5890
0.75	0.5889	0.5887	0.5885	0.5883	0.5882	0.5880	0.5878	0.5877	0.5875	0.5873
0.76	0.5871	0.5870	0.5868	0.5866	0.5865	0.5863	0.5861	0.5859	0.5858	0.5856
0.77	0.5854	0.5853	0.5851	0.5839	0.5847	0.5846	0.5844	0.5842	0.5841	0.5839
0.78	0.5837	0.5835	0.5834	0.5832	0.5830	0.5829	0.5827	0.5825	0.5824	0.5822
0.79	0.5820	0.5818	0.5817	0.5815	0.5813	0.5812	0.5810	0.5808	0.5806	0.5805
0.80	0.5803	0.5801	0.5799	0.5797	0.5796	0.5794	0.5792	0.5790	0.5788	0.5786
0.81	0.5785	0.5783	0.5781	0.5779	0.5777	0.5775	0.5773	0.5772	0.5770	0.5768
0.82	0.5766	0.5704	0.5762	0.5760	0.5759	0.5757	0.5755	0.5753	0.5751	0.5749
0.83	0.5748	0.5746	0.5744	0.5742	0.5740	0.5738	0.5736	0.5735	0.5733	0.5731
0.84	0.5729	0.5727	0.5725	0.5723	0.5722	0.5720	0.5718	0.5716	0.5714	0.5712
0.85	0.5711	0.5703	0.5707	0.5705	0.5703	0.5701	0.5699	0.5698	0.5696	0.5694
0.86	0.5692	0.5690	0.5688	0.5686	0.5685	0.5683	0.5681	0.5679	0.5677	0.5675
0.87	0.5674	0.5672	0.5670	0.5668	0.5666	0.5664	0.5662	0.5661	0.5659	0.5657
0.88	0.5655	0.5653	0.5651	0.5650	0.5648	0.5646	0.5644	0.5642	0.5640	0.5638
0.89	0.5637	0.5635	0.5633	0.5631	0.5629	0.5627	0.5625	0.5624	0.5622	0.5620
0.90	0.5618	0.5616	0.5614	0.5612	0.5610	0.5608	0.5606	0.5604	0.5602	0.5600
0.91	0.5598	0.5596	0.5594	0.5592	0.5590	0.5588	0.5586	0.5584	0.5582	0.5580
0.92	0.5578	0.5576	0.5574	0.5572	0.5570	0.5568	0.5565	0.5563	0.5561	0.5559
0.93	0.5557	0.5555	0.5553	0.5551	0.5549	0.5547	0.5545	0.5543	0.5541	0.5539
0.94	0.5537	0.5535	0.5533	0.5531	0.5529	0.5527	0.5525	0.5523	0.5521	0.5519
0.95	0.5517	0.5515	0.5513	0.5511	0.5509	0.5508	0.5505	0.5503	0.5501	0.5499
0.96	0.5497	0.5495	0.5493	0.5491	0.5489	0.5487	0.5485	0.5483	0.5481	0.5479
0.97	0.5477	0.5475	0.5473	0.5471	0.5469	0.5467	0.5464	0.5462	0.5460	0.5458
0.98	0.5456	0.5454	0.5452	0.5450	0.5448	0.5446	0.5444	0.5442	0.5440	0.5438
0.99	0.5436	0.5434	0.5432	0.5430	0.5428	0.5426	0.5424	0.5422	0.5420	0.5418
1.00	0.5426	0.5414	0.5412	0.5409	0.5407	0.5405	0.5403	0.5400	0.5398	0.5396
1.01	0.5394	0.5391	0.5389	0.5387	0.5385	0.5383	0.5380	0.5378	0.5376	0.5374
1.02	0.5371	0.5369	0.5367	0.5365	0.5362	0.5360	0.5358	0.5356	0.5354	0.5351
1.03	0.5349	0.5347	0.5345	0.5344	0.5340	0.5338	0.5336	0.5333	0.5331	0.5329
1.04	0.5327	0.5325	0.5322	0.5320	0.5318	0.5316	0.5313	0.5311	0.5309	0.5307
1.05	0.5305	0.5302	0.5300	0.5298	0.5296	0.5293	0.5291	0.5289	0.5287	0.5228
1.06	0.5282	0.5280	0.5278	0.5276	0.5273	0.5271	0.5269	0.5267	0.5264	0.5262
1.07	0.5260	0.5258	0.5255	0.5253	0.5251	0.5249	0.5247	0.5244	0.5242	0.5240

M	0	1	2	3	4	5	6	7	8	9
1.08	0.5238	0.5235	0.5233	0.5231	0.5229	0.5229	0.5224	0.5222	0.5219	0.5217
1.09	0.5215	0.5213	0.5211	0.5209	0.5206	0.5204	0.5202	0.5200	0.5197	0.5195
1.10	0.5193	0.5190	0.5188	0.5185	0.5183	0.5189	0.5178	0.5175	0.5173	0.5170
1.11	0.5168	0.5165	0.5163	0.5160	0.5158	0.5155	0.5153	0.5150	0.5148	0.5145
1.12	0.5143	0.5140	0.5137	0.5135	0.5132	0.5130	0.5127	0.5125	0.5122	0.5120
1.13	0.5118	0.5115	0.5113	0.5110	0.5108	0.5105	0.5103	0.5100	0.5098	0.5095
1.14	0.5093	0.5090	0.5088	0.5085	0.5083	0.5080	0.5078	0.5075	0.5073	0.5070
1.15	0.5068	0.5065	0.5062	0.5060	0.5057	0.5055	0.5052	0.5050	0.5047	0.5045
1.16	0.5042	0.5040	0.5037	0.5035	0.5032	0.5030	0.5027	0.5025	0.5022	0.5020
1.17	0.5017	0.5015	0.5012	0.5010	0.5007	0.5005	0.5002	0.5000	0.4997	0.4995
1.18	0.4992	0.4990	0.4987	0.4985	0.4982	0.4980	0.4977	0.4975	0.4972	0.4970
1.19	0.4967	0.4965	0.4962	0.4960	0.4957	0.4955	0.4952	0.4950	0.4947	0.4945
1.20	0.4942	0.4939	0.4936	0.4933	0.4930	0.4928	0.4925	0.4922	0.4912	0.4916
1.21	0.4913	0.4910	0.4907	0.4904	0.4901	0.4899	0.4896	0.4893	0.4890	0.4887
1.22	0.4884	0.4881	0.4878	0.4875	0.4872	0.4870	0.4867	0.4864	0.4861	0.4858
1.23	0.4855	0.4852	0.4849	0.4846	0.4843	0.4841	0.4838	0.4835	0.4832	0.4829
1.24	0.4826	0.4823	0.4820	0.4817	0.4814	0.4812	0.4809	0.4806	0.4803	0.4800
1.25	0.4797	0.4794	0.4791	0.4788	0.4785	0.4783	0.4780	0.4777	0.4774	0.4771
1.26	0.4768	0.4765	0.4762	0.4759	0.4756	0.4754	0.4751	0.4748	0.4745	0.4742
1.27	0.4739	0.4736	0.4733	0.4730	0.4727	0.4725	0.4722	0.4719	0.4712	0.4713
1.28	0.4710	0.4707	0.4704	0.4701	0.4698	0.4696	0.4693	0.4690	0.4687	0.4684
1.29	0.4681	0.4678	0.4675	0.4672	0.4669	0.4667	0.4664	0.4661	0.4658	0.4655
1.30	0.4652	0.4649	0.4645	0.4642	0.4638	0.4635	0.4631	0.4628	0.4625	0.4621
1.31	0.4618	0.4616	0.4611	0.4607	0.4604	0.4601	0.4597	0.4594	0.4590	0.4586
1.32	0.4583	0.4580	0.4577	0.4573	0.4570	0.4566	0.4563	0.4559	0.4556	0.4553
1.33	0.4549	0.4546	0.4542	0.4539	0.4535	0.4532	0.4529	0.4525	0.4522	0.4518
1.34	0.4515	0.4511	0.4508	0.4505	0.4501	0.4498	0.4494	0.4491	0.4487	0.4484
1.35	0.4481	0.4477	0.4474	0.4470	0.4467	0.4463	0.4460	0.4457	0.4453	0.4450
1.36	0.4446	0.4443	0.4439	0.4436	0.4432	0.4429	0.4426	0.4422	0.4419	0.4415
1.37	0.4412	0.4408	0.4405	0.4402	0.4398	0.4395	0.4391	0.4388	0.4384	0.4381
1.38	0.4378	0.4374	0.4371	0.4367	0.4364	0.4360	0.4357	0.4354	0.4350	0.4347
1.39	0.4343	0.4340	0.4336	0.4333	0.4330	0.4326	0.4323	0.4319	0.4316	0.4312
1.40	0.4309	0.4305	0.4301	0.4296	0.4292	0.4288	0.4284	0.4280	0.4275	0.4271
1.41	0.4267	0.4263	0.4258	0.4254	0.4250	0.4264	0.4242	0.4237	0.4233	0.4229
1.42	0.4225	0.4221	0.4216	0.4212	0.4208	0.4204	0.4200	0.4195	0.4191	0.4187
1.43	0.4183	0.4178	0.4174	0.4170	0.4166	0.4162	0.4157	0.4153	0.4149	0.4145
1.44	0.4141	0.4136	0.4132	0.4128	0.4124	0.4120	0.4115	0.4111	0.4107	0.4103
1.45	0.4099	0.4094	0.4090	0.4086	0.4082	0.4077	0.4073	0.4069	0.4065	0.4061
1.46	0.4056	0.4052	0.4048	0.4044	0.4040	0.4035	0.4031	0.4027	0.4023	0.4018
1.47	0.4014	0.4010	0.4005	0.4001	0.3997	0.3993	0.3989	0.3985	0.3980	0.3976
1.48	0.3972	0.3968	0.3964	0.3959	0.3955	0.3951	0.3947	0.3943	0.3938	0.3934
1.49	0.3930	0.3926	0.3921	0.3917	0.3913	0.3909	0.3905	0.3900	0.3896	0.3892
1.50	0.3888	0.3883	0.3878	0.3874	0.3869	0.3864	0.3859	0.3854	0.3850	0.3845
1.51	0.3840	0.3835	0.3830	0.3825	0.3820	0.3816	0.3811	0.3806	0.3801	0.3796
1.52	0.3791	0.3786	0.3781	0.3776	0.3771	0.3766	0.3760	0.3755	0.3750	0.3745
1.53	0.3740	0.3735	0.3730	0.3724	0.3719	0.3714	0.3709	0.3704	0.3698	0.3693

M	0	1	2	3	4	5	6	7	8	9
1.54	0.3688	0.3683	0.3677	0.3672	0.3667	0.3662	0.3656	0.3651	0.3646	0.3640
1.55	0.3635	0.3630	0.3624	0.3619	0.3613	0.3608	0.3602	0.3597	0.3591	0.3586
1.56	0.3580	0.3574	0.3569	0.3663	0.3557	0.3552	0.3546	0.3540	0.3534	0.3528
1.57	0.3523	0.3517	0.3511	0.3506	0.3500	0.3494	0.3488	0.3482	0.3477	0.3471
1.58	0.3465	0.3459	0.3453	0.3447	0.3441	0.3435	0.3429	0.3423	0.3417	0.3411
1.59	0.3405	0.3399	0.3393	0.3386	0.3380	0.3374	0.3368	0.3362	0.3355	0.3349

Figure A.14
Values for slope method

row for the first two decimals of M and the column corresponding to the third decimal place value of M. For example, referring to the first row of the table, if the value of M is 0.400, the value of R is 0.6432. For a value of M equal to 0.401, the value of R will be 0.6431. The product of R and the distance between the current leads gives a distance value Dp.

The second potential lead is now moved to the distance Dp from the ground electrode under testing and the measurement repeated. The value thus obtained is the resistance of ground electrode under test.

Appendix B

IEE system classification based on grounding practices

B.1 Common neutral grounding practices in low-voltage consumer installations as per UK Code BS: 7671: 2000 (IEE wiring regulations)

British Standard BS: 7671: 2000 (IEE wiring regulations) discusses the grounding of low-voltage installations in detail and has provided a method of classifying supply systems based on the type of grounding adopted as well as the method used to extend the system ground to consumer installations. The standard also discusses the comparative merits of the different types of systems for specific applications. The salient features are summarized here.

B.1.1 3-Letter classification code

Low-voltage systems supplying to consumer premises are predominantly solidly grounded. Protective ground connection to consumer premises (or extending the supply system ground to consumer premises) is however done in different ways. The common system categories are defined below using a 3-letter classification (based on IEE Standards).

Note that in these descriptions, 'system includes' both the supply and the consumer installation, and 'live parts' include the neutral conductor.

First letter

T The live parts in the system have one or more direct connections to ground.
I The live parts in the system have no connection to ground or are connected only through a high impedance.

Second letter

T All exposed metal parts/enclosures of electrical equipment are connected to the ground conductor which is then connected to a local ground electrode.
N All exposed metal parts/enclosures of electrical equipment are connected to the ground conductor which is then connected to the ground provided by the supply system.

Remaining letter(s)

C Combined neutral and protective ground functions (same conductor).

S Separate neutral and protective ground functions (separate conductors).

B.1.2 Common types of systems

TN system

A system having one or more points of the source directly grounded with the exposed metal parts being connected to that point by protective conductors. It is further subdivided into the following types depending on the neutral-ground connection configuration.

TN-C system

A system in which the same conductor functions as the neutral and protective conductor throughout the supply and consumer installation (refer Figure B.1).

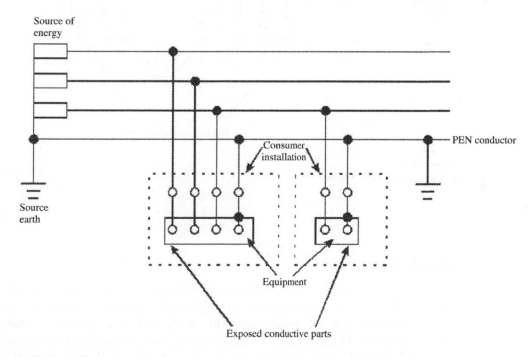

Figure B.1
Schematic of a TN-C system

TN-S system

A system in which separate conductors are provided for neutral and protective ground functions throughout the system. In this type of system, the utility provides a separate ground conductor back to the substation. This is most commonly done by having a grounding clamp connected to the sheath of the supply cable which provides a connection to the ground conductor of the supply side and the grounding terminal of the consumer installation (refer Figure B.2).

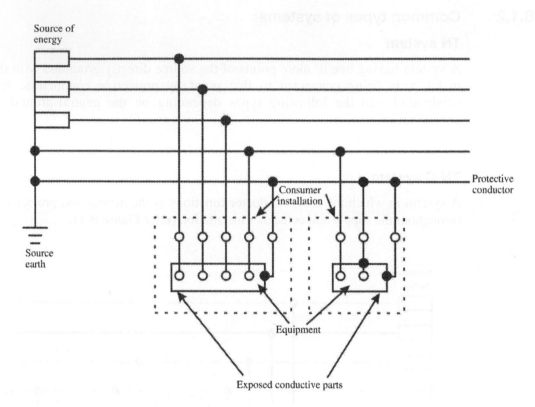

Figure B.2
Schematic of a TN-S system

TN-C-S system

A system in which the neutral and protective functions are done by a single conductor in a part of the system. In this system, in supply side neutral and ground are combined, but they are separated in the installation. This is also called as protective multiple earthing (PME for short). The grounding terminal of the consumer installation is connected to the supplier's neutral. Any breakage of the common neutral cum ground wire, called sometimes as PEN (protective earth and neutral) conductor, can result in the enclosures of electrical equipment inside the premises assuming line voltage when there is insulation failure. It is therefore essential to maintain the connection integrity of this common neutral-cum-ground conductor (refer Figure B.3).

TT   No ground provided by supplier; installation requires own ground rod (common with overhead supply lines) (refer Figure B.4).

IT   Supply is, for example, portable generator with no ground connection, installation supplies own ground rod (refer Figure B.5).

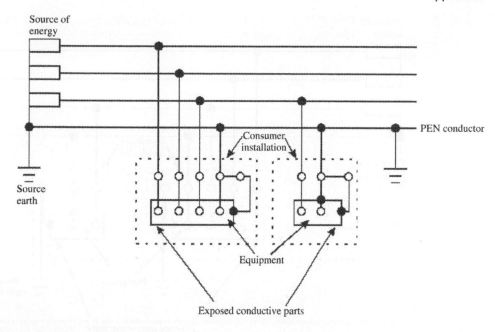

Figure B.3
Schematic of a TN-C-S system

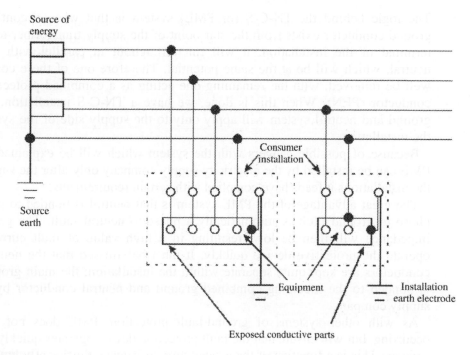

Figure B.4
Schematic of a TT system

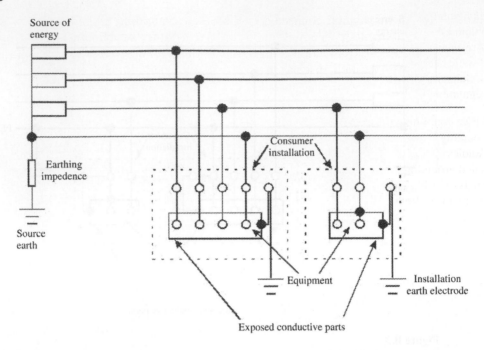

Figure B.5
Schematic of an IT system

B.2 More on TN-C-S systems

The logic behind the TN-C-S (or PME) system is that when a continuous metallic ground conductor exists from the star point of the supply transformer to the grounding terminal of the installation, it will run throughout in parallel with the installation neutral, which will be at the same potential. Therefore one of these conductors can as well be removed, with the remaining one acting as a combined protective and neutral conductor (PEN). When this is done, we have a TN-C-S installation. The combined ground and neutral system will apply only to the supply side of the system and not to the installation.

Because of possible dangers with the system which will be explained further below, PME can be installed by the electricity supply company only after the supply system and the installations it feeds have complied with certain requirements.

The great advantage of the PME system is that neutral is bonded to ground so that a phase to ground fault is automatically a phase to neutral fault. The ground-fault loop impedance will then be low, resulting in a high value of fault current, which will operate the protective device quickly. It must be stressed that the neutral and ground conductors are kept quite separate within the installation; the main grounding terminal is bonded to the incoming combined ground and neutral conductor by the electricity supply company.

As with other systems of ground-fault protection, PME does not prevent a fault occurring, but will ensure that the fault protection device operates quickly when that fault appears. This is a function of the ground loop resistance. Sufficiently low loop resistance will cause high currents but can be cleared faster. A lower current could result in a longer duration of fault because of which the energy dissipation through the point of fault can be quite high and could easily start a fire.

An installation connected to a protective multiple ground supply is subject to special requirements concerning the size of grounding and bonding leads, which are generally larger in cross-section than those for installations fed by supplies with other types of grounding. Full discussions with the electricity supply company are necessary before commencing such an installation to ensure that their needs will be satisfied. The cross-sectional area of the equipotential bonding conductor is related to that of the neutral conductor as shown in Figure B.6.

Neutral Conductor Section (mm^2)	Main Equipotential Bonding Conductor Section (mm^2)
35 or less	10
Over 35 and up to 50	16
Over 50 and up to 95	25

Figure B.6
Minimum cross-sectional area of main equipotential bonding conductor for PME-fed installations

B.3   Installations where use of TN-C-S system is prohibited

Danger can arise when the non-current-carrying metalwork of an installation is connected to the neutral, as is the case with a PME-fed system. The ground system is effectively in parallel with the neutral and can thus share the normal neutral current under certain conditions. This current will not only be that of the installation itself, but may also be part of the neutral current of neighboring installations. It therefore follows that the grounding wires of an installation may carry a significant current even when the main supply to that installation is switched off. This could clearly cause a hazard if in a potentially explosive part of an installation, such as a petrol storage tank, the ground wire were to carry part of the neutral current of a number of installations. For this reason, the use of PME supply system is prohibited in the case of petrol-filling stations. Such installations must be fed from TN-S supply systems.

The difficulty of ensuring that bonding requirements are met on construction sites means that PME supplies must not be used for temporary supplies. Electricity Supply Regulations also do not permit the use of PME supplies to feed caravans and caravan sites. We will discuss the requirements for these consumers briefly below.

B.3.1   Construction sites

The objective of electrical installation on a construction site is to provide lighting and power to enable the work to proceed. By the very nature of the installation, it will be subjected to the kind of rough treatment, which is unlikely to be applied to most fixed installations. Those working on the site may sometimes be ankle deep in mud and thus are particularly susceptible to an electric shock. They may be using portable tools in situations where danger is more likely than in most factory situations. Special regulations are applicable to construction power installations to minimize the danger to working personnel. Apart from the use of TN-S system certain additional requirements are also applicable.

These are as follows:

- Distribution and supply equipment must be protected to IP44. This means provision of mechanical protection from objects more than 1 mm thick and protection from splashing water. Such equipment will include switches and isolators to control circuits and to isolate the incoming supply. The main isolator must be capable of being locked or otherwise secured in the 'off' position. Emergency switches should disconnect all live conductors including the neutral.
- Sockets on a construction site must be separated extra-low voltage (SELV) or protected by a residual current circuit breaker (RCD) with an operating current of not more than 30 mA, or must be electrically separate from the rest of the supply, each socket being fed by its own individual transformer. The SELV is unlikely for most applications because 12 V power tools would draw too much current to be practical. Most sockets are likely to be fed at 110 V from center-tapped transformers and so will comply with this requirement.
- Cables and their connections must not be subjected to strain, and cables must not be run across roads or walkways without mechanical protection. Circuits supplying equipment must be fed from a distribution assembly including over-current protection, a local RCD if necessary and socket outlets where needed. Socket outlets must be enclosed in distribution assemblies, fixed to the outside of the assembly enclosure or fixed to a vertical wall. Sockets must not be left unattached, as is often the case on construction sites.
- Such installations are also by their very nature temporary. As the construction proceeds they will be moved and altered. It is usual for such installations to be subjected to thorough inspection and testing at intervals, which will never exceed 3 months.
- The equipment used must be suitable for the particular supply to which it is connected, and for the duty it will meet on site. Where more than one voltage is in use, plugs and sockets must be non-interchangeable to prevent misconnection. Six levels of voltage are recognized for a construction site installation.

They are:

- 25-V single-phase SELV for portable hand-lamps in damp and confined situations
- 50-V single-phase, center-point grounded for hand lamps in damp and confined situations
- 400-V three-phase, for use with fixed or portable equipment with a load of more than 3750 W
- 230-V single-phase, for site buildings and fixed lighting
- 110-V three-phase, for portable equipment with a load up to 3750 W
- 110-V single-phase, fed from a transformer, often with a grounded center-tapped secondary winding, to feed transportable tools and equipment, such as floodlighting, with a load of up to 2 kW. This supply ensures that the voltage to ground should never exceed 55 V. The primary winding of the transformer must be RCD protected unless the equipment fed is to be used indoors.

The requirements for construction sites will also apply to:

- Sites where repairs, alterations or additions are carried out
- Demolition of buildings

- Public engineering works
- Civil engineering operations, such as road building, coastal protection, etc.

The special requirements for construction sites do not apply to temporary buildings erected for the use of the construction workers, such as offices, toilets, cloakrooms, dormitories, canteens, meeting rooms, etc. These areas/buildings are not subject to changes as construction work progresses and are thus exempt from these requirements.

B.3.2   Marina supply systems

Another typical situation where PME systems are not used is the marina power supply system. A marina is a location, often a harbor, for leisure craft to berth. Like residential caravans, such craft require external power supplies, and this section is intended to ensure the safety and standardization of such supplies. The marina will often include shore-based facilities such as offices, workshops, toilets, leisure accommodation, and so on and is exempt from these special requirements.

The electrical installation of a marina is subject to hazards not usually encountered elsewhere. These include the presence of water and salt, the movement of the craft, increased corrosion due to the presence of salt water and dissimilar metals and the possibility of equipment being submerged due to unusual wave activity in bad weather.

The neutral of a PME system must not be connected to the grounded system of a boat so that the hazards, which follow the loss of continuity in the supply PEN conductor, are avoided. This rules out the use of PME supplies for marinas. Where this is the supply provided, it must be converted to a TT system at the main distribution board by provision of a separate ground electrode system of driven rods or buried mats with no overlap of resistance area with any ground associated with the PME supply. If the marina is large enough, it may be that the supply company will provide a separate transformer and a TN-S system.

A common installation method is to provide a feed from the shore to a floating pontoon via a bridge or ramp, and then to equip the pontoon with socket outlets to feed the craft moored to it. Socket outlets may be single or three phase. Where multiple single-phase sockets are installed on the same pontoon, they must all be connected to the same phase of the supply unless fed through isolating transformers. Socket outlets should be positioned as close as possible to the berth of the vessel they feed, with a minimum of one socket per berth, although up to six sockets may be provided in a single enclosure. Each socket outlet must be provided with a means of isolation which breaks all poles on TT systems and must be protected by an over-current device such as a fuse or a circuit breaker. Groups of socket outlets must be RCD protected. Each socket or group of sockets must be provided with a durable and legible notice giving instructions for the electricity supply.

B.3.3   Caravan power supply

A caravan is a leisure accommodation vehicle, which reaches its site by being towed by a vehicle. A motor caravan is used for the same purpose, but has an engine, which allows it to be driven; the accommodation module on a motor caravan may sometimes be removed from the chassis. Caravans will often contain a bath or a shower, and in these cases the special requirements for such installations will apply. Railway rolling stock is not included in the definition as a caravan.

All dangers associated with fixed electrical installations are also present in and around caravans. Added to these are the problems of moving the caravan, including connection and disconnection to and from the supply, often by unskilled people. Grounding is of prime importance because the dangers of shock are greater. For example, the loss of the main protective conductor and a fault to the metalwork in the caravan is likely to go unnoticed until someone makes contact with the caravan whilst standing outside it. The requirements of the Electricity Supply Regulations do not allow the supply neutral to be connected to any metalwork in a caravan, which means that PME supplies must not be used to supply them.

The power supply to a caravan must be made using approved type of coupler at a height of not more than 1.8 m from ground. The coupler socket (fixed to the caravan body) must have a spring-loaded lid, which will protect the socket when caravan is traveling. A clear notice must be affixed near the socket to indicate voltage, current and frequency of the supply required by the caravan. This socket must be connected to a main isolator with RCD protection with an operating current of 30 mA, which on operation will disconnect all live conductors. All metal parts of the caravan, with the exception of metal sheets forming part of the structure, must be bonded together and to a circuit-protective conductor, which must not be smaller than 4 mm^2 except where it forms part of a sheathed cable or is enclosed in conduit.

B.3.4   Other locations where PME supplies are not permitted

There are locations such as swimming pools, saunas, hospitals, etc., where danger to human beings is more probable. The reason for this is that in some of the locations, the human beings are partly unclothed, without shoes and are in contact with water or their body is wet. The effect of any contact with live electrical parts could therefore be particularly dangerous.

A hospital where open-heart surgery is performed poses even greater danger because of the effect of stray electric currents finding their way to the human heart through medical appliances used in the treatment. Also explosive hazards may exist in certain hospital locations handling anesthetic gases.

The use of electrical systems in these cases will depend on the zone of use. These zones are defined in each case depending on the presence of vulnerable human body in the vicinity. Use of lower-voltage isolated systems, ground fault alarms and RCD is advised depending on the application.

Appendix C

IEEE exposure classifications

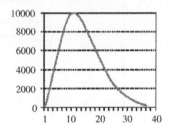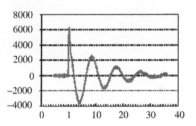

Figure 1
Combination wave and ring wave

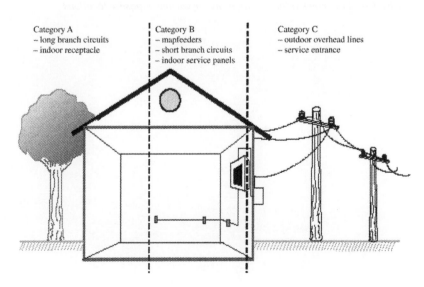

Figure 2
IEEE C62.41 location categories

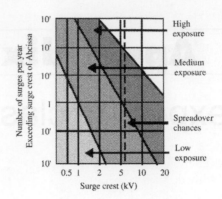

Figure 3
IEEE C62.41 exposure levels

Category	Level	Voltage (kV)	0.5 μs × 100 kHz Ring wave current (A)	1.2 × 50 μs (V) 8 × 20 μs (A) Combination wave current (kA)
A1	Low	2	70	
A2	Medium	3	130	
A3	High	6	200	
B1	Low	2	170	1
B2	Medium	4		2
B3	High	6	30	3
			500	
C1	Low	6		3
C2	Medium	10		5
C3	High	20		10

Figure 4
IEEE C62.41 current/voltage waveforms for various exposure locations

Appendix D

Glossary of terms related to grounding

D.1 Glossary of terms

This glossary of terms is an extract from IEEE Std 1100 – 1992 'IEEE Recommended Practice for Powering and Grounding Sensitive Electronic Equipment (IEEE Emerald Book)'.

D.2 Introduction

The electronic power community is pervaded by terms that have no scientific definition; one of the purposes of this chapter is to eliminate the use of those words. Another purpose of this chapter is to define those terms that will aid in the understanding of concepts within this recommended practice. Wherever possible, definitions were obtained from the IEEE Std 100 – 1988 [4]. The second choice was to use other appropriate sources, and the final choice was to use a new definition that conveys a common understanding for the word as used in the context of this recommended practice.

This appendix is divided into three parts. First, an alphabetical listing of definitions is provided in D.3. The reader is referred to the IEEE Std 100 – 1988 for all words not listed herein. Section D.4 lists those terms that have been deliberately avoided in this document because of no generally accepted single technical definition. These words find common use in discussing AC distribution-related power problems, but tend not to convey significant technical meaning. Section D.5 lists abbreviations that are employed throughout IEEE recommended practices.

D.3 Alphabetical listing of terms used in this recommended practice

The primary source for the definitions in this section is IEEE Std 100 – 1988 [4]. This section does not include any device or equipment definitions (for example, isolation transformers, uninterruptible power systems); the reader is advised to refer to the index.

Bonding   (1) The electrical interconnecting of conductive parts, designed to maintain a common electrical potential [2]. (2) The permanent joining of metallic parts to form an electrically conductive path, which will assure electrical continuity and the capacity to conduct safely any current likely to be imposed [2].

Commercial power   Electrical power furnished by the electric power utility company.

Common-mode noise   The noise voltage that appears equally and in phase from each current-carrying conductor to ground.

Note: For the purposes of this recommended practice, this definition expands the existing definition in IEEE Std 100 – 1988 [4] (previously given only for signal cables) to the power conductors supplying sensitive electronic equipment.

Coupling   Circuit element or elements, or network that may be considered common to the input mesh and the output mesh and through which energy may be transferred from one to the other [4].

Crest factor   Ratio between the peak value (crest) and rms value of a periodic waveform [4].

Critical load   Devices and equipment whose failure to operate satisfactorily jeopardizes the health or safety of personnel, and/or results in loss of function, financial loss or damage to property deemed critical by the user.

Degradation failure   See Failure, degradation.

Differential-mode noise   See Noise, Transverse mode.

Direct-reading ammeters   Ammeters that are employed with a series shunt and that carry some of the line current through them for measurement purposes – they are part of the circuit being measured.

Displacement power factor   See Power factor, displacement.

Distortion factor   The ratio of the root mean square of the harmonic content to the root mean square value of the fundamental quantity, expressed as a percent of the fundamental [4] – also referred to as total harmonic distortion [6].

Dropout   A loss of equipment operation (discrete data signals) due to noise, sag or interruption.

Dropout voltage   The voltage at which a device will release to its de-energized position (for this recommended practice, the voltage at which a device fails to operate).

Efficiency (of a power system)   The relationship between the input power that a power system draws and the corresponding power that it is able to supply to the load (kilowatt out/kilowatt in).

Equipment grounding conductor   The conductor used to connect the non-current-carrying parts of conduits, raceways and equipment enclosures to the grounding electrode at the service equipment (main panel) or secondary of a separately derived system (for example, isolation transformer) – (this term is defined more specifically in the NEC [2], Section 100).

Failure, degradation   Failure that is both gradual and partial. *Note*:   In time, such failure may develop into complete failure [4].

Failure mode   The effect by which a failure is observed [4].

Flicker   A variation of input voltage sufficient in duration to allow visual observation of a change in electric light source intensity.

Form factor (periodic function)   The ratio of the root mean square value to the average absolute value, averaged over a full period of the function [4].

Forward transfer impedance   An attribute similar to internal impedance, but at frequencies other than the nominal (for example, 60 Hz power frequency) – knowledge of the forward transfer impedance allows the designer to assess the capability of the power source to provide load current (at the harmonic frequencies) needed to preserve a good output voltage waveform – generally, the frequency range of interest is 60 Hz– 3 kHz, for 5–60 Hz power systems and 20–25 kHz for 380–480 Hz power systems.

Frequency deviation   An increase or decrease in the power frequency – the duration of a frequency deviation can be from several cycles to several hours.

Ground   A conducting connection, whether intentional or accidental, by which an electric circuit or equipment is connected to the earth, or to some conducting body of relatively large extent that serves in place of the earth.

Note: It is used for establishing and maintaining the potential of the earth (or of the conducting body) or approximately that potential, on conductors connected to it, and for conducting ground currents to and from each (or the conducting body) [4].

Ground electrode   A conductor or group of conductors in intimate contact with the earth for the purpose of providing a connection with the ground [2].

Ground electrode, concrete encased   A grounding electrode completely encased within concrete, located within and near the bottom of a concrete foundation or footing or pad, that is in direct contact with the earth. (This term is defined more specifically in Article 250 of the NEC [2].)

Ground grid   A system of interconnected base conductors arranged in a pattern over a specified area and buried below the surface of the earth – the primary purpose of the ground grid is to provide safety for workmen by limiting potential differences within its perimeter to safe levels in case of high currents, which could flow if the circuit being worked became energized for any reason or if an adjacent energized circuit faulted – metallic surface mats and gratings are sometimes utilized for the same purpose [4] – this term should not be used when referring to a signal reference structure, which is defined in this chapter.

Ground, high-frequency reference   See Signal reference structure.

Ground impedance tester   A multifunctional instrument designed to detect certain types of wiring and grounding problems in low-voltage power distribution systems.

Ground loops   A potentially detrimental loop formed when two or more points in an electrical system that are nominally at ground potential area connected by a conducting path such that either or both points are not at the same ground potential [4].

Ground, radial   A conductor connection by which separate electrical circuits or equipment are connected to earth at one point – sometimes referred to as a star ground.

Ground, Ufer   See Ground electrode, concrete encased.

Ground window   The area through which all grounding conductors, including metallic raceways, enter a specific area – it is often used in communications systems through which the building grounding system is connected to an area that would otherwise have no grounding connection.

Harmonic distortion   The mathematical representation of the distortion of the pure sine waveform – see Distortion factor.

Impulse   See Transient.

Input power factor (of a system)   Specifies the ratio of input kilowatts to input kilovoltamperes at rated or specified voltage and load.

Input voltage range (of a power system)   The range of input voltage that the system can operate over.

Inrush   The amount of current that a load draws when it is first turned on.

Interruption   The complete loss of voltage for a time period.

Isolated equipment ground   An insulated equipment grounding conductor run in the same conduit or raceway as the supply conductors – this conductor is insulated from the metallic raceway and all ground points throughout its length. It originates at an isolated ground type receptacle or equipment input terminal block and terminates at the point where neutral and ground are bonded at the power source (this term is defined more specifically in the NEC [2], Sections 250–74 and 250–75).

Isolation   Separation of one section of a system from undesired influences of other sections.

Linear load   An electrical load device which, in steady-state operation, presents an essentially constant load impedance to the power source throughout the cycle of applied voltage.

Noise   Electrical noise is unwanted electrical signals that produce undesirable effects in the circuits of the control systems in which they occur [4] (for this recommended practice, 'control systems' is intended to include sensitive electronic equipment in total or in part).

Noise, common mode   See Common-mode noise.

Noise, differential mode   See Transverse-mode noise.

Noise, normal mode   See Transverse-mode noise.

Noise, transverse mode   See Transverse-mode noise.

Non-linear load   Electrical load that draws current discontinuously or whose impedance varies during the cycle of the input AC voltage waveform.

Non-linear load current   Load current that is discontinuous or is not proportional to the AC voltage.

Notch   A switching (or other) disturbance of the normal power voltage waveform, lasting less than a half-cycle, which is initially of opposite polarity than the waveform and is thus subtractive from the normal waveform in terms of the peak value of the disturbance voltage – this includes complete loss of voltage for up to a half-cycle – see Transient.

Outage   See Interruption.

Output (reverse transfer) impedance (of a power source)   Similar to forward transfer impedance, but it describes the characteristic impedance of the power source as seen from the load, looking back at the source – see Forward transfer impedance.

Over-voltage   An rms increase in the AC voltage, at the power frequency, for durations greater than a few seconds – see Swell and surge.

Phase shift   The displacement in time of one periodic waveform relative to other waveform(s).

Power disturbance   Any deviation from the nominal value (or from some selected thresholds based on load tolerance) of the input AC power characteristics.

Power disturbance monitor I   Instrumentation developed specifically for the analysis of voltage and current measurements.

Power factor, displacement   The displacement component of power factor; the ratio of the active power of the fundamental wave, in watts, to the apparent power on the fundamental wave, in volt-amperes [4].

Power factor, total   The ratio of the total power input in watts to the total volt-ampere input [4].

Power quality   The concept of powering and grounding sensitive electronic equipment in a manner that is suitable to the operation of that equipment.

Radial ground   See Ground, radial.

Receptacle circuit tester   A device that, by a pattern of lights, is intended to indicate wiring errors in receptacles – receptacle circuit testers have some limitations – they may indicate incorrect wiring, but cannot be relied upon to indicate correct wiring.

Recovery time   Time interval needed for the output voltage or current to return to a value within the regulation specification after a step load or line change [2] – also may indicate the time interval required to bring a system back to its operating condition after an interruption or dropout.

Recovery voltage   The voltage that occurs across the terminals of a pole of a circuit interrupting device upon interruption of the current [4].

Safety ground   See Equipment grounding conductor.

Sag   An rms reduction in the AC voltage, at the power frequency, for durations from a half-cycle to a few seconds – see Notch, under-voltage. *Note*:   The IEC terminology is dip.

Shield   As normally applied to instrumentation cables, a conductive sheath (usually metallic) applied over the insulation of a conductor or conductors, for the purpose of providing means to reduce coupling between the conductors so shielded and other conductors that may be susceptible to, or that may be generating, unwanted electrostatic or electromagnetic fields (noise).

Shielding   Is the use of a conducting barrier between a potentially disturbing noise source and sensitive circuitry – shields are used to protect cables (data and power) and electronic circuits – they may be in the form of metal barriers, enclosures or wrappings around source circuits and receiving circuits.

Signal reference structure   A system of conductive paths among interconnected equipment that reduces noise-induced voltages to levels that minimize improper operation – common configurations include grids and planes.

Slew rate   Rate of change of (AC voltage) frequency.

Star ground   See Ground, radial.

Star-connected circuit   A polyphase circuit in which all the current paths of the circuit extend from a terminal of entry to a common terminal or conductor (which may be the neutral conductor). *Note*:   In a three-phase system, this is sometimes called a Y (or wye) connection [4].

Surge   See Transient.

Surge reference equaliser   A surge-protective device used for connecting equipment to external systems whereby all conductors connected to the protected load are routed, physically and electrically, through a single enclosure with a shared reference point between the input and output ports of each system.

Swell   An rms increase in the AC voltage, at the power frequency, for durations from a half-cycle to a few seconds – see Over-voltage and Surge.

Transfer time (uninterruptible power supply)   The time that it takes an uninterruptible power supply to transfer the critical load from the output of the inverter to the alternate source or back again.

Transient   A sub-cycle disturbance in the AC waveform that is evidenced by a sharp brief discontinuity of the waveform –   may be of either polarity and may be additive to or subtractive from the nominal waveform – see Swell, Notch and Over-voltage.

Transverse-mode noise   (with reference to load device input AC power) – noise signals measurable between or among active circuit conductors feeding the subject load, but not between the equipment grounding conductor or associated signal reference structure and the active circuit conductors.

Unbalanced load regulation   A specification that defines the maximum voltage difference between the three output phases that will occur when the loads on the three are of different levels.

Under-voltage   An rms decrease in the AC voltage, at the power frequency, for durations greater than a few seconds.

Voltage distortion   Any deviation from the nominal sine waveform of the AC line voltage.

Voltage regulation   The degree of control or stability of the rms voltage at the load – often specified in relation to other parameters, such as input-voltage changes, load changes or temperature changes.

D.4   Words avoided because of no single technical definition

The following words have a varied history of usage, and some may have specific definitions for other applications. It is an objective of this recommended practice that the following ambiguous words not be used to generally describe problem areas or solutions associated with the powering and grounding of sensitive electronic equipment.

- Blackout
- Brownout
- Clean ground
- Clean power
- Computer grade ground
- Conducting barriers
- Counterpoise ground
- Dedicated ground
- Dirty ground
- Dirty power
- Equipment safety grounding conductor
- Frame ground
- Frequency shift
- Glitch
- Natural electrodes
- Power surge
- Raw power
- Raw utility power
- Shared circuits
- Shared ground
- Spike
- Sub-cycle outages
- Type I, II, III power disturbances.

D.5   Abbreviations and acronyms

The following abbreviations are utilized throughout this recommended practice:

AFD	adjustable frequency drive
ALVRT	automatic line voltage regulating transformer
ASAI	average service availability index
CATV	cable accessed television
COTC	central office trunk cable
CPC	vcomputer power center
CPU	central processing unit
CRT	cathode ray tube
CVT	constant voltage transformer
EFT	electrical fast transient
EGC	equipment grounding conductor
EMC	electromagnetic compatibility
EMI	electromagnetic interference
EMT	electrical metallic tubing
ESD	electrostatic discharge
FMC	flexible metal conduit
HF	high frequency
IEC	International Electrotechnical Commission
IG	isolated/insulated grounding

IMC	intermediate metal conduit
IT	isolation transformer
LDC	line drop compensator
M-G	motor-alternator/generator
MCT	metal cable tray
MTBF	mean time between failures
NEMP	nuclear electromagnetic pulse
NEC	National Electrical Code
NIST	National Institute of Standards and Technology
OEM	original equipment manufacturer
OSHA	Occupational Safety and Health Administration
PC	personal computer
PDU	power distribution unit
PLC	power line conditioner
PWM	pulse width modulation
RFI	radio frequency interference
RMC	rigid metal conduit
SDS	separately derived AC system
SE	service entrance
SG	solidly grounded; solid grounding (see equipment grounding conductor)
SI	solidly interconnected
SRG	signal reference grid
SRP	signal reference plane
SRS	signal reference structure
THD	total harmonic distortion
TVSS	transient voltage surge suppressor
UL	Underwriters Laboratories
UPS	uninterruptible power supply
VDT	video display terminal

D.6   References

[1]   ANSI C84.1 – 1989, American National Standard for Electric Power Systems and Equipment – Voltage Ratings (60 Hz).

[2]   ANZI/NFPA 70 – 1993, National Electrical Code.

[3]   IEEE Std C62 – 1990, Complete 1990 edition: Guides and Standards for Surge Protection.

[4]   IEEE Std 100 – 1988, IEEE Standard Dictionary of Electrical and Electronics Terms (ANSI).

[5]   IEEE Std 142 – 1991, IEEE Recommended Practice for Grounding of Industrial and Commercial Power Systems (IEEE Green Book).

[6]   IEEE Std 519 – 1992, IEEE Guide for Harmonic Control and Reactive Compensation of Static Power Converters (ANSI).

[7]   P1159, Recommended Practice on Monitoring Electric Power Quality, D2/June 30, 1992.

Appendix E
Steps to ensure effective substation grounding

E.1 Substation grounding

An electrical substation is a critical resource in a power system. Safe operation of a substation calls for a properly designed and installed grounding system. A well-designed grounding system will ensure reliable performance of the substation over its entire service life. In the main manual, the means of ensuring safety of personnel by proper grounding practices has been highlighted in different chapters. This appendix provides them in one place for ready reference.

How does good grounding improve substation reliability? Good grounding path of sufficiently low impedance ensures fast clearing of faults. A fault remaining in the system for long may cause several problems including those of power system stability. Faster clearing thus improves overall reliability. It also ensures safety. A ground fault in equipment causes the metallic enclosure potential to rise above the 'true' ground potential. An improper grounding results in a higher potential and also results in delayed clearing of the fault (due to insufficient current flow). This combination is essentially unsafe because any person coming into contact with the enclosure is exposed to higher potentials for a longer duration. Therefore, substation reliability and safety must be as 'built-in' as possible by good grounding scheme, which in turn will ensure faster fault clearing and low enclosure potential rise.

E.2 Ensuring proper grounding

The following steps, when put into practice, will ensure a reliable, safe and trouble-free substation grounding system.

E.2.1 Size conductors for anticipated faults

Conductors must be large enough to handle any anticipated faults without fusing (melting). Failure to use proper fault time in design calculations creates a high risk of melted conductors. Two aspects govern the choice of conductor size: the first is the fault current that will flow through the conductor and the second is the time for which it can flow. The fault current depends on the impedance of the ground fault loop. The time of current flow is decided by the setting of the protective relays/circuit-breaking devices, which will operate to clear the fault. The IEEE 80 suggests using a time of 3.0 s for the design of small substations. This time is also equal to the short-time rating of most switchgear.

E.2.2   Use the right connections

It is very evident that the connections between conductors and the main grid and between the grid and ground rods are as important as the conductors themselves in maintaining a permanent low-resistance path to ground. The basic issues here are:

- The type of bond used for the connection of the conductor in its run, with the ground grid and with the ground rod
- The temperature limits, which a joint can withstand.

The most frequently used grounding connections are mechanical pressure type (which will include bolted, compression and wedge-type construction) and exothermically welded type. Pressure-type connections produce a mechanical bond between conductor and connector by means of a tightened bolt-nut or by crimping using hydraulic or mechanical pressure. This connection either holds the conductors in place or squeezes them together, providing surface-to-surface contact with the exposed conductor strands. On the other hand, the exothermic process fuses the conductor ends together to form a molecular bond between all strands of the conductor.

Temperature limits are stated in standards such as IEEE 80 and IEEE 837 for different types of joints based on the joint resistance normally obtainable with each type. Exceeding these temperatures during flow of fault currents may result in damage to the joint and cause the joint resistance to increase, which will result in further overheating. The joint will ultimately fail and result in grounding system degradation or total loss of ground reference with disastrous results.

E.2.3   Ground rod selection

In MV and HV substations, where the source and load are connected through long overhead lines, it often happens that the ground fault current has no metallic path and has to flow through the groundmass (earth). This means that the ground rods of both source and load side substations have to carry this current to or from the groundmass. The ground rod system should be adequate to carry this current and ground resistance of the grounding system assumes importance.

The length, number and placement of ground rods affect the resistance of the path to earth. Doubling of ground rod length reduces resistance by a value of 45%, under uniform soil conditions. Usually, soil conditions are not uniform and it is vital to obtain accurate data by measuring ground rod resistance with appropriate instruments.

For maximum efficiency, grounding rods should be placed no closer together than the length of the rod. Normally, this is 10 ft (3 m). Each rod forms an electromagnetic shell around it, and when the rods are too close, the ground currents of the shells interfere with each other. It should be noted that as the number of rods is increased, the reduction of ground resistance is not in inverse proportion. Twenty rods do not result in 1/20th of the resistance of a single rod but only reduce it by a factor of 1/10th.

For economic reasons, there is a limit to the maximum distance between rods. Normally, this limit is taken as 6 m. At more than 6 m, the cost of additional conductor needed to connect the rods makes the design economically attractive. In certain cases, the substation layout may not have the required space and acquiring the needed space may involve substantial expense. Four interconnected rods on 30 m centers will reduce resistivity 94% over one rod but require at least 120 m of conductor. On the other hand, four rods placed 6 m apart will reduce resistivity 81% over one rod and use only 24 m of conductor.

E.2.4 Soil preparation

Soil resistivity is an important consideration in substation grounding system design. The lower the resistivity, the easier it is to get a good ground resistance. Areas of high soil resistivity and those with ground frost (which in turn causes the soil resistivity to increase by orders of magnitude) need special consideration. The highest ground resistivity during the annual weather cycle should form the basis of the design since the same soil will have much higher resistivity during dry weather when percentage of moisture in the ground becomes very low. One approach to take care of this problem is to use deep driven ground rods so that they are in contact with the soil zone deep enough to remain unaffected by surface climate. The other approach is to treat the soil around the ground rod with chemical substances that have the capacity to absorb atmospheric/soil moisture. Use of chemical rods is one such solution.

E.2.5 Attention to step and touch potentials

Limiting step and touch potential to safe values in a substation is vital to personnel safety. Step potential is the voltage difference between a person's feet and is caused by the voltage gradient in the soil at the point where a fault enters the earth. The potential gradient is steepest near the fault location and thereafter reduces gradually. Just 75 cm away from the entry point, voltage usually will have been reduced by 50%. Thus at a point of 75 cm from the fault (which is less than the distance of a normal step), a fatal potential of a few kilovolts can exist.

Touch potential represents the same basic hazard, except the potential exists between the person's hand and his or her feet. This happens when a person standing on the ground touches a structure of the substation, which is conducting the fault current into ground (for example, when an insulator fixed on a gantry flashes over, the gantry dissipates the current to earth). Since the likely current path within the human body runs through the arm and heart region instead of through the lower extremities, the danger of injury or death is far greater in this case. For this reason, the safe limit of touch potential is usually much lower than that of step potential.

In both situations, the potential can essentially be greatly reduced by an equipotential wire mesh safety mat installed just below ground level. This mesh will have to be installed in the immediate vicinity of any switches or equipment a worker might touch, and connected to the main ground grid. Such an equipotential mesh will equalize the voltage along the worker's path and between the equipment and his or her feet. With the voltage difference (potential) thus essentially eliminated, the safety of personnel is virtually guaranteed.

An equipotential wire mesh safety mat is usually fabricated from #6 or #8 AWG copper or copper-clad wire to form a 0.5×0.5 m or 0.5×1 m mesh. Many other mesh sizes are available. To ensure continuity across the mesh, all wire crossings are brazed with a 35% silver alloy. Interconnections between sections of mesh and between the mesh and the main grounding grid should be made so as to provide a permanent low-resistance high-integrity connection.

E.2.6 Grounding using building foundations

Concrete foundations below ground level provide an excellent means of obtaining a low-resistance ground electrode system. Since concrete has a resistivity of about 30Ω m. at $20 \,°C$, a rod embedded within a concrete encasement gives a very low electrode resistance compared to most rods buried in the ground directly. Since buildings are usually constructed using steel-reinforced concrete, it is possible to use the reinforcement rod as the conductor of the electrode by ensuring that an electrical connection can be established with the main rebar of each foundation. The size of the rebar as well as the bonding between the bars of different concrete members must be done so as to ensure that ground fault currents can be handled without excessive heating. Such heating may cause weakening and eventual failure of the concrete member itself. Alternatively, copper rods embedded within concrete can also be used.

The use of 'Ufer' grounds (named after the person who was instrumental in the development of this type of grounding practice) has significantly increased in recent years. Ufer grounds utilize the concrete foundation of a structure plus building steel as a grounding electrode. Even if the anchor bolts are not directly connected to the reinforcing bars (rebar), their close proximity and the conductive nature of concrete will provide an electrical path.

There are a couple of issues to be considered while planning for grounding using the foundations as electrodes. A high fault current (lightning surge or heavy ground fault) can cause moisture in the concrete to evaporate suddenly to steam. This steam, whose volume is about 1800 times of its original volume when existing as liquid, produces forces that may crack or otherwise damage the concrete.

The other factor has to do with ground leakage currents. The presence of even a small amount of DC current will cause corrosion of the rebar. Because corroded steel swells to about twice its original volume, it can cause extremely large forces within the concrete. Although AC leakage will not cause corrosion, the earth will rectify a small percentage of the AC to DC. In situations where the anchor bolts are not bonded to the rebar, concrete can disintegrate in the current path.

Damage to concrete can be minimized either by limiting the duration of fault current flow (by suitable sensitive and fast acting protective devices) or by providing a metallic path from the rebar through the concrete to an external electrode. That external electrode must be sized and connected to protect the concrete's integrity.

Proper design of Ufer grounds provides for connections between all steel members in the foundation and one or more metallic paths to an external ground rod or main ground grid. Excellent joining products are available in the market, which are especially designed for joining rebars throughout the construction. By proper joining of the rebars, exceptionally good performance can be achieved. An extremely low-resistance path to earth for lightning and earth fault currents is ensured as the mass of the building keeps the foundation in good contact with the soil.

E.2.7   Grounding the substation fence

Metallic fences of substations should be considered just as other substation structures. The reason for this is that the overhead HV lines entering or leaving a substation may snap and fall on the fence. Unless the fence is integrated with the rest of the substation grounding system, a dangerous situation may develop. Persons or livestock in contact with the fence may receive dangerous electric shocks.

Utilities vary in their fence-grounding specifications, with most specifying that each gate post and corner post, plus every second or third line post, be grounded. All gates should be bonded to the gate posts using flexible jumpers. All gate posts should be interconnected. In the gate swing area, an equipotential wire mesh safety mat can further reduce hazards from step and touch potentials when opening or closing the gate.

It is recommended that the fence ground should be tied into the main ground grid, as it will reduce both grid resistance and grid voltage rise. Internal and perimeter gradients must be kept within safe limits because the fence is also at full potential rise. This can be accomplished by extending the mesh with a buried perimeter conductor that is about 1 m outside the fence and bonding the fence and the conductor together at close intervals (so that a person or grazing animal touching the fence will stand on the equipotential surface so created).

E.2.8 Special attention to operating points

To protect the operator in case of a fault, it should be ensured that he is not subjected to high touch or step potentials when a fault happens in the equipment he is operating. This calls for use of a safety mesh close to these operating points on which the operator will stand and operate the equipment. There are four types of safety mats.

1. A steel grate or plate on supporting insulators. This works only if the operator can be kept completely isolated on the grate. Therefore, insulators must be kept clean. Any vegetation in the vicinity should be cut or eliminated completely (this approach is similar to the insulating rubber mats placed in front of most indoor electrical equipment). Safety is ensured by increasing the resistance of current path, so that the current flowing through the operator's body into the ground does not exceed safe values.
2. A steel grate on the surface, permanently attached to the grounded structure. This arrangement has the operator standing directly on the grate.
3. Bare conductor buried (in a coil or zig-zag pattern) under the handle area and bonded to the grounded structure.
4. Prefabricated equipotential wire mesh safety mat buried under the handle area and bonded to the grounded structure. This is likely to be the least expensive choice.

In all but the first arrangement, both the switch operating handle and the personnel safety grate (or mat) should be exothermically welded to structural steel, thus ensuring nearly zero voltage drop.

E.2.9 Surge arrestors must be grounded properly

When there is a surge in the electrical system (by indirect lightning strikes or due to switching) surge arrestors placed near all critical equipment divert surge energy to ground and protect the equipment from being subjected to the surges. Usually, surges involve a very fast rise time during which the current changes from zero to extremely high values of several kiloamperes. It is therefore necessary that the conducting path from the grounding terminal of the surge arrestor to the earth must have minimum impedance. Even a small amount of self-inductance offered by a grounding conductor will mean very high impedance because of the steep wavefront of the surge and very high voltages from appearing in the grounding system (albeit briefly). To dissipate the surge current with minimum voltage drop, each surge arrestor ground lead should have a short direct path to earth and should be free of sharp bends (bends act like a coil and increase the inductance). Often surge arrestors are mounted directly on the tank of transformers, close to the HV terminal bushings. In these cases, the transformer tanks and related structures act as the grounding path. It must be ensured that multiple and secure paths to ground are available (this includes making effective connections). Whenever there is any question about the adequacy of these paths, it is recommended to use a separate copper conductor between the arrestor and the ground terminal (or main grounding grid). Since steel structures (due to their multiple members) have lower impedance than a single copper conductor, the grounding conductors should preferably be interconnected to the structure near the arrestor.

E.2.10   Grounding of cable trays

The NEC vide Art. 318 specifies the requirements for cable trays and their use in reducing the induced voltages during a ground fault. All metallic tray sections must be bonded together with proper conducting interconnections. The mechanical splice plates by themselves may not provide an adequate and a reliable ground path for fault currents. Therefore, the bonding jumpers (either the welded type used on steel trays or the lug type) must be placed across each spliced tray joint. If a metallic tray comes with a continuous grounding conductor, the conductor can be bonded inside or outside the tray. When cable tray covers are used, they should be bonded to the tray with a flexible conductor. The trays should also be bonded to the building steel (usually at every other column).

E.2.11   Temporary grounding of normally energized parts

When personnel work on high-voltage electric structures or equipment, any conductive bodies should be grounded as a measure of safety. This is done so that in the event of the circuit becoming live due to inadvertent switching, the safety of personnel (in contact with the parts, which would become live) is ensured. The usual grounding method is to attach a flexible insulated copper cable with a ground clamp or lug on each end. These flexible jumpers require periodic inspection and maintenance.

For cable connections to clamps, welded terminations (either a welded plain stud or a threaded silicon bronze stud welded to the conductor end) will provide a secure, permanent connection. The clamp or lug is solidly connected to ground, then the other clamp is attached to the cable being grounded.

Appendix F

Course exercises

Problem 1

The questions given in this annex are drawn from the principles covered in the course topics and can be answered using the figures, formulae and the tables given in different chapters. In case there are any assumptions not stated explicitly in the course, help is given in the form of hints.

An ungrounded 5 kV, 3-wire system has the following cable feeders.

Cable Size	Length (m)	Capacitance (Micro-Farads per km)
3c × 240 Sq.mm	550	0.43
3c × 150 Sq.mm	2600	0.35
3c × 120 sq.mm	1240	0.32
3c × 50 Sq.mm	780	0.23

Calculate the ground fault current of this system. Calculate the inductance of the grounding reactor for

(a) Exactly balancing the capacitive current and
(b) For obtaining a net current value of 5 A inductive.

Hint: The fault current is the total of charging currents of all three phases added numerically.

Problem 2

A 380 V circuit with a ground fault current of 6 kA is protected by a fuse of 250 A with a let through $(I^2 \times t)$ value of $600\,000\,\text{As}^2$. Find out the tolerable touch potential value for a human of 70 kg weight neglecting the Resistance of ground underneath the feet and the mutual ground resistance between the feet. A resistance of $1000\,\Omega$ for the human body to current flow can be assumed.

Problem 3

A circuit has a ground fault current of 3000 A. Calculate the cross section of the copper grounding wire to be used assuming hat the current will be interrupted in 300 m sec. The final temperature that the grounding wire can assume should be restricted to 150 °C. Initial temperature may be assumed to be 70 °C.

Problem 4

A mast of 60 m height has a 10 m high building adjacent to it. If the farthest point of the base of the building is at a distance of 25 m from the base of the mast verify whether the building is completely enveloped within the attraction sphere of the mast. Assume two cases with peak lightning current values of

- 5 kA
- 20 kA.

Problem 5

A transmission line with two circuits has a shield wire for lightning protection. The height of the shield wire from ground is 18 m. The three phase conductors of each circuit are arranged on either side of the pole at an offset of 2 m from the center. The vertical clearance between the top wire and the shield wire is 3 m. The phase conductors are arranged vertically with 1.5 m clearance. Verify whether the phase conductors fall within the attraction sphere of the shield wire for peak lightning currents of

- 5 kA
- 20 kA.

Problem 6

A motor with a belt drive is operating in an environment with ignitable dust. The capacitance value of the belt can be assumed as 50 pF. The minimum energy of a spark to cause ignition in this environment is 120 mJ. Verify whether there is any danger of ignition due to static buildup.

Hint: Use voltage build up figure given in Table 5.1.

Problem 7

A test using the 4-pin method is being conducted to determine the soil resistivity. The test data are as follows.

Spike length	150 cm
Spike spacing	10 m soil
temperature during test	30 °C
C Soil moisture during test	24%
Resistance value	2.4 Ω

Calculate the soil resistivity under test conditions. Also calculate the resistivity at soil moisture content of 14% and soil temperature of 10 °C.

Hint: Consider that the soil type as typical top soil. Use tables in Figures. 6.4 and 6.5.

Problem 8

A ground rod of 40-mm diameter is driven into soil to a depth of 250 cm. Calculate the resistance of the soil surrounding this rod at distances of 10 cm, 15 cm, 20 cm, 40 cm, 60 cm, 100 cm, 200 cm, 250 cm, 300 cm, 500 cm and 800 cm from the rod. Soil resistivity can be taken as 100 Ω m. Express the values as % of the resistance at 800 cm.

Hint: Assume soil shells (slices) of thickness 5 cm up to 20 cm distance; 10 cm between 20 and 100 cm distance and 50 cm after 100 cm distance. Area may be calculated using the inner face of the shell. Make a tabular form for recording the resistance of successive shells and calculate cumulative resistance values as you go along.

Problem 9

A driven rod electrode of made of 40-mm diameter GI pipe is buried to a depth of 3 m in a soil of resistivity 50 Ω m. Find the resistance of the ground rod, In order to improve the resistance 12 such rods are used in a rectangular array and are bonded together. Calculate the overall ground resistance of this array.

Problem 10

Calculate for the above example the ground fault current that can be safely dissipated into the ground for the following fault clearance times.

- 0.1 s
- 0.5 s
- 1.0 s
- 5.0 s.

Appendix G

Answers

Problem 1

This problem uses vector diagrams shown in Figure 2.5.

Step 1: Calculation of per phase capacitances

For 3c × 240 Sq.mm cable

$$C = 0.43 \times 550/1000$$
$$= 0.2365 \text{ Micro Farads}$$

For 3c × 150 Sq.mm cable

$$C = 0.35 \times 2600/1000$$
$$= 0.91 \text{ Micro Farads}$$

For 3c × 120 sq.mm cable

$$C = 0.32 \times 1240/1000$$
$$= 0.3968 \text{ Micro Farads}$$

For 3c × 50 Sq.mm cable

$$C = 0.23 \times 780/1000$$
$$= 0.1794 \text{ Micro Farads}$$

(Above calculation uses the fact that capacitance value is proportional to the length.)

Step 2: Total capacitance per phase for the entire system

$$C_{\text{Total}} = 1.7227 \text{ Micro Farads or } (1.7727 \times 10^{-6}) \text{ Farads}$$

(When a number of capacitors are in parallel the effective capacitance is the sum of individual capacitances.)

Step 3:   Capacitive reactance per phase

$$Xc = \frac{1}{2 \cdot \pi \cdot F \cdot C_{\text{Total}}}$$

$$= \frac{1}{2 \cdot \pi \cdot 60 \cdot 1.7727 \times 10^{-6}}$$

$$= 1539.8\,\Omega$$

Note: *F* is taken as 60 Hz

Step 4:   Charging current per phase

$$Ic = \frac{\text{Line Voltage}}{Xc \cdot 3^{0.5}}$$

$$= \frac{5000}{1.732 \times Xc)}$$

$$= 1.875\,\text{A}$$

Step 5:   Total Capacitive fault current

$$Ic = 3 \times Ic$$

$$= 5.625\,\text{A}$$

Step 6:   Inductive reactance for case (a)

If *X* is the inductive reactance connected to neutral, inductive current flowing through *X* is given by:

$$I_R = \frac{5000}{(1.732 \times X)}$$

X for obtaining a value of inductive current 5.625 A is calculated by substitution.

$$X = 513.2153\,\Omega$$

Inductance *L* can be calculated using

$$X = 2 \cdot \pi \cdot F \cdot L$$

$$L = 1.36\,\text{Henries}$$

Step 7:   Inductive reactance for case (b)

In this case the net fault current value is a sum of *Ic* and 5 A

$$I_R = 5 + 5.625\,\text{A}$$

$$= 10.625\,\text{A}$$

Value of X and L required to obtain this current can be calculated using the same equations in Step 6 as

$$X = 271.7 \, \Omega$$

$$L = 0.721 \text{ Henries}$$

Problem 2

Basis of solution is tolerable touch potential value in 3.2

Step 1:   Time of fault current flow before interruption

The let through value of fuse ($I^2 \cdot Ts$) is given as 600,000 As^2. Since the current value is given as 6000 A, Time Ts can be calculated as

$$Ts = \frac{I^2 \cdot Ts}{6000 \times 6000}$$

$$= \frac{600000}{6000 \times 6000}$$

$$= 0.01 \, \text{s}$$

Step 2:   Tolerable body current

Tolerable body current I_B for Time Ts can be calculated using the equation

$$I_B = \frac{0.157}{\sqrt{Ts}}$$

$$= 1.214 \, \text{A}$$

Step 3:   Tolerable touch potential

Using the relation Vtouch = $R_B \times I_B$ Where body resistance R_B can be taken as 1000 Ω.

Tolerable value of touch potential for the duration Ts can be calculated as 1214 V.

Note

In case you are wondering how this value is possible in a 380 V Circuit: The calculation only shows that this value of voltage can be tolerated by a human body when the fault is cleared in the time shown. It establishes that the system is safe since the touch voltage in a 380 V system cannot reach this value.

Problem 3

Basis of solution is the formula for temperature limits of grounding wire given in Appendix A Section A.1.

Given that:

Max permissible temperature is 150 °C

Operating temperature is 70 °C

Ground fault current is 3000 A

Fault duration is 300 ms or 0.3 s

Substituting in the formula

$$\frac{I^2 \cdot t}{A} = 0.0297 \cdot \ln \frac{[Tm + 234]}{Ti + 234}$$

$$\text{Cross sectional area } A = \frac{3000 \cdot 3000 \cdot 0.3}{0.2097 \cdot \log \frac{[Tm + 234]}{Ti + 234}}$$

$$= 389 \times 10^6 \text{ Circular Mils}$$

Problem 4

To be solved using the attraction radius formula in Chapter 4

Step 1: Calculating the distance from mast top to base of building

Mast height is given as 60 m

Horizontal distance to base of building is 25 m

Distance L from top of mast to base of farthest point of the building is calculated as:

$$L = \sqrt{60^2 + 25^2}$$
$$= 65 \text{ m}$$

Step 2: Attraction radius

Attraction radius can be calculated using the formula

$$R_A = 0.84 \times h^{0.6} \times I^{0.74}$$

For case 1 (5 kA stroke)

$$I = 5000 \text{ A}$$

Substituting

$$Ra = 32.2 \text{ m}$$

For case ss2 (20 kA stroke)

$$I = 20000 \text{ A}$$

Substituting

$$Ra = 89.9 \text{ m}$$

Result

The attractive sphere of the mast does not envelope the building fully for a stroke of peak 5 KA.

The attractive sphere of the mast envelopes the building fully for a stroke of peak 20 KA.

Problem 5

To be solved using the attraction radius formula in Chapter 4 for conductors.

In this case, the following formula will be applied to calculate the zone of protection afforded by the shield wire.

$$R_D = 0.67 \times h^{0.6} \times I^{0.74}$$

Where

R_D is the radial distance of attraction on either side of the conductor in meters

h is the height of the conductor from ground in meters

I is the peak current of lightning in kA.

The protection available from the shield wire and the relative disposition of the line conductors is shown in Figure G.1.

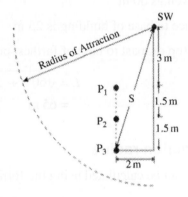

SW : Shield Wire
P1, P2, P3: Line Conductors in Vertical Formation

Figure G.1
Disposition of conductors

We can infer that the line is fully protected if:

$$R_D > S$$

Where S is the distance between the shield wire and the lowermost line conductor.
From the figure it can be seen that:

$$S = \sqrt{(6^2 + 2^2)}$$
$$= 6.32 \text{ m}$$

For 5 kA peak current

$$R_D = 0.67 \times 18^{0.6} \times 5000^{0.74}$$
$$= 12.48 \text{ m}$$

For 20 kA peak current

$$R_D = 0.67 \times 18^{0.6} \times 20000^{0.74}$$
$$= 34.8 \text{ m}$$

The line conductors are protected for both values of peak lightning current.

Problem 6

To be solved using the spark energy formula given in Chapter 5.

Step 1: Calculation of spark energy

Given that:

Capacitance value of the belt is 50 pF

Voltage buildup taken from Figure 5.1 for belted drives is 60 to 100 kV

Taking a maximum value of 100 kV for conservative design

Substituting in formula Spark Energy mJ $0.5 \times C \times V^2 \times 10^{-9}$

$$\text{Energy} = 0.5 \times 50 \times 10 \times (100 \times 10^3)^2$$
$$= 250 \text{ mJ}$$

Step 2: Compare with energy to cause ignition given as 120 mJ

Conclusion

The energy of accumulated charge in belt drive is adequate to cause ignition.

Problem 7

To be solved using the formula in Chapter 6 for calculation of soil resistivity

Step 1: Calculate resistivity under given soil conditions

Given that:

Spike spacing S is 10 m

Resistance R is 2.4 Ω

Soil temp. is 30 °C

Moisture 24%

Resistivity is given by substitution

$$\rho = 2\pi S R$$
$$= 150.8 \ \Omega \text{ m}$$

Step 2: Correction for soil moisture in Table given in Figure 6.2

For top soil, Soil resistivity with 24% moisture is 100 Ω m

Soil resistivity with 14% moisture is 250 Ω m

Correction factor is 2.5 for correcting test value at 24% moisture to 14% moisture condition.

Resistivity corrected for Moisture is 2.5 × 150.8

$$= 377 \ \Omega \, m$$

Step 3:   Correction for soil temperature based on table given in Figure 6.3

For top soil, Soil resistivity at 30°C is 60 Ω m

At 10°C soil resistivity is 80 Ω m

Correction factor is 1.33 for correcting from test condition at 30°C to actual at 10°C.

Resistivity corrected for Moisture is 1.33 × 377

$$= 501.4 \ \Omega \, m$$

Problem 8

To be solved using the principle illustrated in Figure 6.6 and the formula for this condition

Given that:

Ground rod length is 2.5 m

Soil resistivity is 100 Ω m

Diameter of electrode is 40 mm (4 cm)

Step 1:   Divide the soil around the electrode into a number of slices

The first slice is from the periphery of rod to 5 cm from the center

The next slice is 5 to 10 cm and so on

Step 2:   The inner surface area

This can be calculated as the sum of the concentric cylindrical surface and the hemispherical bottom surface.

Cylindrical surface area for the first slice can be calculated using the formula

$$\text{Area} = 2 \times \pi \times \text{Radius of the cylinder} \times \text{Length}$$

Substituting

$$A1 = \frac{2 \times \pi \times 2.5 \times 2}{100}$$

Hemispherical area can be calculated using the formula

$$\text{Area} = 2 \times \pi \times \text{Radius} \times \text{Radius}$$

$$A2 = \frac{2 \times \pi \times 2 \times 2}{100 \times 100}$$

Total surface area is the sum of A1 and A2

$$A = 0.317 \text{ sq. m}$$

Resistance of the shell can be calculated using the formula

$$R = \frac{\rho L}{A}$$

Where

   R is the Resistance of the shell in Ohms
   L is the thickness of the shell in meters
   A is the inner surface area of the shell in sq. meters
   And ρ is the soil resistivity in ohm meters (100 in this case)

Substituting these values

$$R = \frac{100 \times 0.03}{0.317}$$
$$= 9.4637224 \ \Omega.$$

Step 3

Repeat this for subsequent slices and tabulate the results.

Step 4

The slice thickness given in the hint under the problem may be used for calculation.

Step 5

Calculate cumulative resistance using this table.

Step 6

Express the cumulative resistance as % of resistance calculated for the soil mass up to a radius of 8 m

Note

It would be of interest to compare the values obtained with the values given in Table 6.7

The values calculated using a spreadsheet program are shown below and can be compared with manually obtained values.

Radial Distance (cm)	Surface Area (sq.m)	Resistance of Slice (Ω)	Cum. Resistance (Ω)	% of Cum. Res to Value at 8 m
2				
5	0.32	9.47	9.47	24.80
10	0.80	6.24	15.71	41.15
15	1.63	3.06	18.77	49.16
20	2.50	2.00	20.77	54.40
30	3.39	2.95	23.72	62.12
40	5.28	1.89	25.61	67.08
50	7.29	1.37	26.98	70.67
60	9.43	1.06	28.04	73.45
70	11.69	0.86	28.90	75.69
80	14.08	0.71	29.61	77.55
90	16.59	0.60	30.21	79.13
100	19.23	0.52	30.73	80.49
150	22.00	2.27	33.00	86.44
200	37.71	1.33	34.33	89.92
250	56.57	0.88	35.21	92.23
300	78.57	0.64	35.85	93.90
350	103.71	0.48	36.33	95.16
400	132.00	0.38	36.71	96.15
450	163.43	0.31	37.01	96.95
500	198.00	0.25	37.27	97.62
550	235.71	0.21	37.48	98.17
600	276.57	0.18	37.66	98.64
650	320.57	0.16	37.82	99.05
700	367.71	0.14	37.95	99.41
750	418.00	0.12	38.07	99.72
800	471.43	0.11	38.18	100.00

Problem 9

Calculation to be done using the formula for resistance of ground electrodes in Section 6.5.

Step 1:   Arrive at Resistance of a single electrode

Given:

Soil resistivity (ρ) is 50 Ω m

Electrode dia (d) is 40 mm (0.04 m)

Length L is 3 m

Substituting in the formula

$$R = \frac{\rho}{2\pi L} \times \{Ln \ (8 \ L/D) \ -1\}$$

where

R is the resistance of the Electrode in Ohms
ρ is the soil resistivity in Ohm meters
L is the length of the buried part of the electrode in meters and
D is the outer diameter of the rod in meters

R can be calculated as 14.31 Ω

Step 2:   Resistance of parallel combination

Given that there are 12 electrodes connected together

The resistance of parallel combination is $R \times F/N$ where $N = 12$ and F can be arrived at from the table given in Figure 6.8

For N = 12 F is 1.8

Substituting

Resistance of the combination is 14.31 $\times$ 1.8/12

Result = 2.147 Ω

Problem 10

To calculate the current carrying capacity for individual electrodes as per formula in Chapter 6 (Section 6.6)

$$I = \frac{34800 \cdot d \cdot L}{\sqrt{\rho \cdot t}}$$

where

I is the maximum permissible current in Amperes
d is the outer diameter of the rod in meters
L is the length of the buried part of the electrode in meters and
ρ is the soil resistivity in Ω meters and
t is the time of the fault current flow in seconds

Given that:

Soil resistivity (ρ) is 50 Ω m

Electrode dia (d) is 40 mm (0.04 m)

Length L is 3 m

Time t has different values 0.1 s 0.5 s 1 s and 5 s.

Substituting in the formula, safe current for a single electrode is:

$$I = \frac{34800 \times 0 \cdot 04 \times 3}{\sqrt{50 \cdot t}}$$

Total current for 12 electrodes can be taken as $12 \times I$

The values can be calculated as

I for 0.1 s = 22410 A

I for 0.5 s = 10022 A

I for 1 s = 7086 A

I for 5 s = 3169 A

Appendix H

Group activities

Instructions

These sheets contain a few of the case studies covered in Chapter 10 but presented in the form of problems.

Participants are requested to form small groups – study the situations outlined in these problems, analyze the facts presented and arrive at probable solutions based on the facts discussed in Chapters 1 through 9.

The instructor may thereafter call for a few of the participants to present their solutions followed by a group discussion. Participants are requested not to refer to Chapter 10 till this discussion is over.

Case study 1

Problem

A steel mill with variable speed drives (VSDs) had problems of frequent tripping of the VSDs with the indication 'over-voltage in AC line'. Each tripping caused severe production disruption and resulted in considerable monetary loss due to lost production.

Other relevant facts

Steady-state measurements by true rms voltmeter showed that voltage was normal and within the specified operating range of the VDS. A power line monitor was then used in the distribution board feeding the VSDs and the incoming power feeder to the mill. At both locations, the monitors showed transient over-voltages of damped oscillatory type waveform with initial amplitude of over 2.0 pu and a ringing frequency of about 700 Hz.

The timing of disturbances coincided with the closing of capacitor banks in the utility substation feeding the steel mill.

It was confirmed by the VSD manufacturer that the VSDs were provided with over-voltage protection set to operate at 1.6 pu voltage for disturbances exceeding 40 µs. Since the switching transients were above this protection threshold, the VSDs tripped (refer Figure H.1).

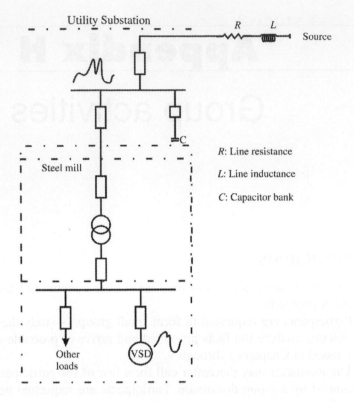

Figure H.1
Distribution arrangement

Analysis

_____

_____

_____

_____

Possible solution

_____

_____

_____

Case study 2

Problem

In an office building with several offices, a constant voltage transformer (CVT) feeding a facsimile machine got overheated one night and started emitting smoke. On getting the fire-alarm signal, the security guard switched off supply to the machine. The following day, the engineer who was investigating his problem noticed that several fluorescent lamps going randomly on and off with an abnormal humming sound from the lamp chokes. This was happening all over the office intermittently as loads were being switched on in the beginning of the workday.

Other relevant facts

The office building was fed by a 500-kVA transformer, which fed a distribution center through a cable with three full capacity line conductors and a half-capacity neutral conductor.

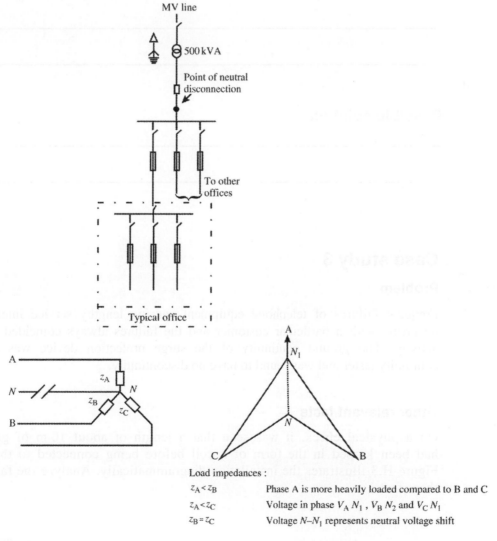

Load impedances :

$z_A < z_B$	Phase A is more heavily loaded compared to B and C
$z_A < z_C$	Voltage in phase $V_A N_1$, $V_B N_2$ and $V_C N_1$
$z_B \approx z_C$	Voltage $N–N_1$ represents neutral voltage shift

Figure H.2
Distribution arrangement

It was found that in the incoming cable termination to the LV distribution center, the neutral lead was red hot and arcing intermittently when the loads were on.

Many offices in the building were installing computer systems fed by UPS equipment in the previous 1 year (non-linear loads).

Refer Figure H.2 and analyze the effect of the neutral failure with a phasor diagram and explain the reason for fluctuations in voltage.

Analysis

_____

_____

_____

_____

_____

Possible solution

_____

_____

Case study 3

Problem

Frequent failures of telephone equipment causing lengthy service interruptions were occurring with a particular customer and the failures always coincided with lightning activity. The ground continuity of the surge protection device was checked by a continuity tester and was found to have no discontinuity.

Other relevant facts

On a physical check, it was seen that a length of about 10 m of grounding wire had been looped in the form of a coil before being connected to the ground rod. Figure H.3 illustrates the installation diagrammatically. Analyze the failure based on these facts.

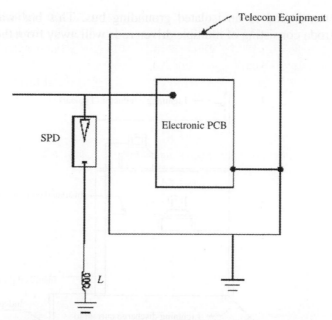

L: Inductance caused by conductor loop in the grounding circuit of LPD

Figure H.3
Connection of communication equipment

Analysis

_____

_____

_____

_____

Possible solution

_____

_____

_____

Case study 4

Problem

In a large office complex, computer errors including system crashes/reboots were happening during thunderstorms. The grounding of computer system had been carried out according to the recommendations of manufacturers. The grounding leads were insulated and terminated to an isolated grounding bus. This bus was connected to a grounding electrode consisting of multiple driven rods well away from the building (refer Figure H.4).

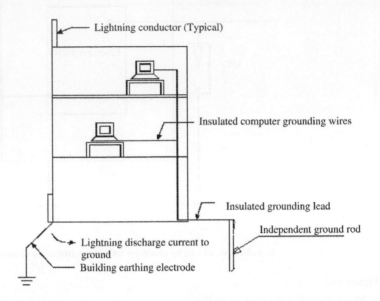

Figure H.4
Computer errors and reboots

Other relevant facts

It was noticed that appreciable voltages were recorded between the building grounding.

Analysis

_____

_____

_____

_____

_____

Possible solution

_____

_____

Case study 5

Problem of monitor display

Problem of wavy monitor display was reported from a new computer installation. Power quality checks did not indicate any abnormality. Subsequently, magnetic field measurements were carried out and indicated presence of high power frequency magnetic fields (refer to Figure H.5 for hint).

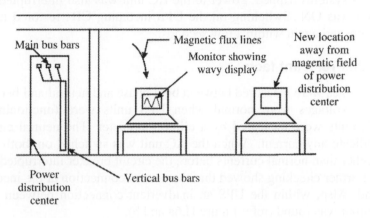

Figure H.5
Display problem in monitor

Analysis

_____

_____

_____

_____

Possible solution

_____

_____

Case study 6

Problem

In an office computer installation, a single-phase distribution board was feeding two 1-kVA UPS systems, each feeding a LAN server. One of the AC units whose power feeder developed a fault was temporarily connected to a spare outlet of the DB feeding the UPS systems. All of a sudden, the circuit breakers on the AC power input of both UPS systems tripped. Power to the AC unit was also interrupted though the feeder to the unit was ON. Switching on the UPS incoming CBs restored normalcy but the tripping happened again within a few seconds.

Other relevant facts

Voltages were measured between both phase and neutral and between neutral and ground. The voltages were normal when UPS units were functioning (AC unit was OFF). Currents were measured by a clip-on ammeter. The neutral circuit of the UPS did not indicate any current. When the AC unit was switched on, both neutral circuits indicated higher than normal currents before the circuit breaker interrupted the current.

Further checking showed that the neutral connection at the incoming line to the DB was bad. Also, within the UPS an inadvertent connection between neutral of AC input and output was found (refer Figure H.6a and b).

Analysis

_____

_____

_____

_____

_____

Possible solution

_____

_____

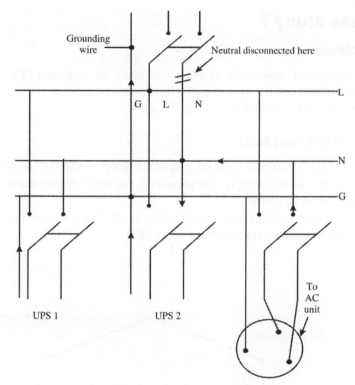

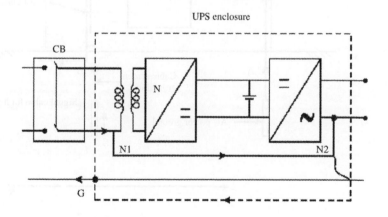

Note: ↓ Arrows show path taken by neutral current of AC unit
due to inadvertent connection between ground and neutral
in the UPS unit (see UPS internal connection shown in Fig b)

(a) Distribution arrangement

(b) UPS internal arrangement

Figure H.6
Problem due to tripping CBs

Case study 7

Problem

A residential consumer faced a problem of repeated TV set failures whenever a thunderstorm hit the area. The TV had a cable connection. No SPDs were provided either on the power supply wires or the TV cable.

Other relevant facts

It was found that two separate grounds were provided in the house at opposite ends. One was by the power supply company and the other by the cable operator. These were not bonded to each other.

   The power ground had a connection to the TV through the power cord and the other through the cable screen (refer Figure H.7).

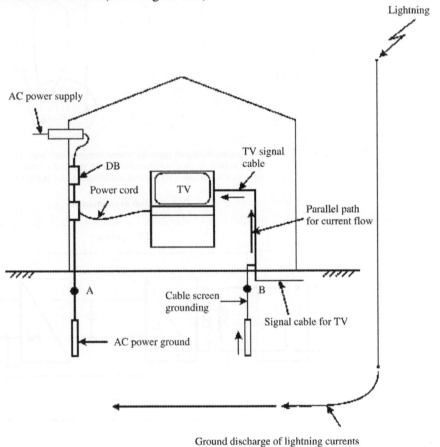

Figure H.7
Failure of TV sets

Analysis

_____

_____

_____

Possible solution

_____

_____

Case study 8

Problem

In an industrial facility, there was a cluster of three buildings each having a process control computer connected with data cabling. Each computer was grounded to the grounding system of the respective building. The grounding systems of the buildings were interconnected through water mains and metallic sheaths of cables, etc. The functioning of the computer systems was very erratic (refer Figure H.8).

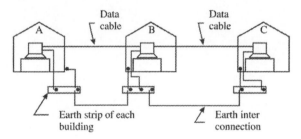

Figure H.8
Facility spread over three buildings

Other relevant facts

It was thought that grounding of the computer installations of the three buildings at a single point in the central building may eliminate erratic operation but it was not taken up because it was a violation of safety codes.

Analysis

_____

_____

_____

_____

_____

Possible solution

_____

_____

Index